Animal Breeding
The Modern Approach

*A Textbook for Consultants, Farmers,
Teachers and for Students
of Animal Breeding*

Published by
Post Graduate Foundation
in Veterinary Science
University of Sydney

Editors

Dr Keith Hammond
Department of Animal Science

Dr Hans-Ulrich Graser
Technical Director
Animal Genetics and Breeding Unit

Mr Alex McDonald
Extension Specialist
Animal Genetics and Breeding Unit

University of New England
Armidale
New South Wales, 2351
Australia

© Copyright 1992 University of Sydney

Post Graduate Foundation in Veterinary Science
P.O. Box A561
Sydney South 2000, NSW, Australia
Tel: (02) 264 2122
        (International) 61 2 264 2122
Fax: (02) 261 4620
        (International) 61 2 261 4620

ISBN 1-875582-20-7

Preface

This book is not so much about the science, but about **how the science can be used in the field** to take advantage of the huge amount of biological variation which exists in our livestock industries. As such, Animal Breeding - The Modern Approach will be of use to all people concerned with achieving genetic change in livestock populations used in meeting human needs for food & fibre, students, advisors and consultants, teachers, administrators and of course farmers themselves.

The book is intended to relate to the range in level of application, viz. the individual herd or flock, groups of herds or flocks, national industries for a number of livestock species, and international initiatives in livestock improvement.

This first issue was developed to commence a series of symposia for which the participants were active field consultants, government and privately employed, and with Animal or Veterinary Science backgrounds. Hence the authors of the sections of this text also delivered the first symposium of the series, and we thank them for both contributions.

The approach taken in the book recognises that successful manipulation of biological variation in modern animal agriculture depends not only on genetic principles and procedures but also on the integration and packaging of these together with a number of other disciplinary areas, including information science and computing, mathematical statistics, economics and what is commonly termed operations research. **Successful applications are based on clear and rather simple principles** and are not complicated and too difficult as is sometimes heard of genetics, **and on the close integration of these principles with particular production systems.**

It is this integration of disciplines and their incorporation with production systems, aimed at efficiently and effectively capitalising on biolological diversity over time, that we term **Animal Breeding**.

The book's design is based on a simple concept termed The Modern Breeding Approach. The book can be read either from start to finish or its section can be used separately.

Keith Hammond
Hans-Ulrich Graser
Alex McDonald
Armidale, September 1992

Acknowledgments

Much of the research and development work by the editors and authors serves as a forerunner to this text and was frequently supported financially by Australia's Rural Industry Research Funds; in particular the

Dairy Research and Development Corporation
Meat Research Corporation
Pig Research and Development Corporation
Wool Research and Development Corporation

The majority of the word processing and typesetting to produce this text was done by Kerry Blair and Gillian Macdougall. The covers graphic was designed by Louise Campbell, and most of the figures were prepared by Don Gentle.

The Editors and Authors

Keith Hammond, BScAgr PhD(Syd), Department of Animal Science, University of New England, Armidale NSW 2351, Australia

Dr Hammond has twenty years experience in Animal Breeding research and application at the national and international levels. He is founding Director of the highly successful Animal Genetics and Breeding Unit, a hybrid research institute of NSW Agriculture and The University of New England. He has played a leading role in the development of genetic evaluation and breeding objective procedures for several livestock industries. He is an author of around 250 technical papers and 3 patents, and has received a number of awards for his contributions to research and industry where he is particularly known for his ability to interpret the science in relation to practical requirements.

Hans-Ulrich Graser, Dr sc agr (Hohenheim), Animal Genetics and Breeding Unit, University of New England, Armidale NSW 2351, Australia

Born in Germany he graduated in 1978 from the University of Hohenheim (Stuttgart) with an emphasis in Animal Breeding. Since then he has been heavily involved in industry research both in Germany and Australia. He has played a major role in developing genetic evaluation procedures for beef and dairy cattle and pig industries. He is Technical Director of AGBU.

Alex McDonald, BAgrSc(LaT), Executive Director, Limousin Society, P.O. Box 262 Armidale, NSW 2350, Australia

Mr McDonald is an Extension Specialist of almost 20 years experience at the regional and national levels, specialising in carcase evaluation and Animal Breeding. He was the National Coordinator for the BREEDPLAN system and responsible for introducing Real Time Ultrasonic Scanning to an industry for use in evaluation of carcase merit on live animals. He is now Executive Director for the Limousin Breed in Australia.

Robert Banks, BAgrSci(Syd) PhD(UNE), Department of Animal Science, University of New England, Armidale NSW 2351, Australia

Dr Banks is the National Co-ordinator of the LAMBPLAN meat sheep evaluation services in Australia, and has played a major role in the design and uptake of the service. Working with NSW Agriculture and the Meat Research Corporation of Australia and located at the Department of Animal Science of the University of New England, he also has under-graduate and graduate teaching responsibilities. Previously, he was responsible for the design and implementation of research and extension activities in performance recording and Animal Breeding at the state level for the dairy cattle, beef cattle, meat sheep and wool sheep industries.

Stephen Barwick, MScAgr(UNE) PhD(Ohio State), Animal Genetics and Breeding Unit, University of New England, Armidale NSW 2351, Australia

Currently leader of NSW Agriculture's Beef Breeding and Evalaution Research Team. From a background of sheep breeding, Dr Barwick joined the AGBU in 1989 where he leads the research and development program of beef breeding objectives and the implementation of these.

Mick Carrick, MScAgr(Syd) PhD(Calif), Queensland Department of Primary Industries, GPO Box 46, Brisbane Qld 4001, Australia

Dr Carrick has worked in the areas of Animal Reproduction and Breeding for 27 years, with experience in North America and Australia and in the development of genetic improvement initiatives for pig, temperate and tropical beef cattle, honeybee and wool sheep industries.

Recently he was Scientific Director and Chief Executive of Merinotech, a group of companies specialising in advanced sheep breeding technologies, before becoming the Chief Geneticist with the Queensland Department of Primary Industries in Australia.

Willi Fuchs, Dr nat techn (Vienna), Markt Neubau, 2880 Kirchberg/WE, Austria

After completing advanced training in Animal Breeding, Dr Fuchs designed and implemented a comprehensive data management system for a large Austrian company before moving to Australia where he worked at both the research and commercial levels designing and implementing the BREEDPLAN International and PIGBLUP genetic evaluation systems, and contributing to the development of the B-OBJECT system. He has experience in dairy cattle, beef cattle and pig industries.

Goddard, Michael, PhD(Melb), Department of Agriculture and Rural Affairs, Division of Dairying, P.O. Box 500, East Melbourne VIC 3002, Australia

Is Senior Geneticist and Head of the Livestock Improvement Unit, Victorian Department of Food and Agriculture. Dr Goddard is involved in research on genetic improvement of dairy and beef cattle and sheep. His background includes research on genetic improvement of beef cattle for tropical Australia; and development of breeding programs for guide dogs.

Brian Kinghorn, PhD (Edin) Dr Agric (Norway), Department of Animal Science, University of New England, Armidale NSW 2351, Australia

Woldwide experience in Animal Breeding and Genetics having held positions in Zimbabwe, Norway, UK and Australia in cattle, sheep and fish industries. Current interests include design, computer simulation and industry application of breeding programs. Since 1988 Dr Kinghorn has been a Director of Merinotech Pty Ltd.

Tom Long, MS (Illinois) PhD (Nebr), Animal Genetics and Breeding Unit, University of New England, Armidale NSW 2351, Australia

Worked in the US swine industry for an international pig breeding company before attaining a PhD in genetic evaluation procedures for swine. Dr Long is currently Co-ordinator of Pig Genetics at ABGU and is involved with improving the PIGBLUP system (micro computer software for genetic evaluation in pigs).

Markus Schneeberger, Dr sc techn (ETH), Schafzuchtverband, Niederönz, CH-3360 Herzogenbuchsee, Switzerland

Dr Schneeberger has 15 years experience in the design of recording and breeding procedures nationally and internationally. He has worked in Europe, North America and Australia, involving the dairy cattle, beef cattle and sheep industries. He has advanced research experience in the use of Mixed-Model and Portfolio theory. After working for several years as a Senior Scientist at AGBU he recently returned to Switzerland to take a leading position with the Swiss Sheep Breeders Association.

Andrew Swan, BRurSc PhD(UNE), CSIRO, Chiswick, Armidale NSW 2350, Australia

Dr Swan is a young scientist with indepth experience in the prospects for and problems with across-breed genetic evaluation procedures. He has experience in both the sheep and beef cattle industries, and is currently a geneticist with Australia's CSIRO, involved in genetic evaluation for merino sire reference schemes, and estimation of genetic parameters and development of breeding objectives for fine and superfine Merino lines.

Contents

PART I: MODERN ANIMAL BREEDING

1 The New Era in Genetic Improvement of Livestock   1
Keith Hammond

The Fundamentals	1
What is Animal Breeding	1
Basic Breeding Biology	2
Does Animal Breeding Work	3
Developments Relevant to Modern Animal Breeding	4
Information in Animal Breeding	7
The New Era in Animal Breeding	9
Taking on New Technology	9
Product Pricing Systems and Genetic Improvement	11
Overview	12

2 The Modern Breeding Approach   13
Keith Hammond

Introduction	13
What are the Key Decision Areas in Animal Breeding?	13
The Three Primary Components of Breeding Operations	16
The Extent of Genetic Differences	18
Which Genetics is Important in Animal Breeding?	19
Some Poorly Understood Concepts in Animal Breeding	23

3 Designing Performance Recording Operations   27
Keith Hammond

Introduction	27
What is Performance Recording	27
What Decisions are Required, and When?	29
What Output is required and When?	31
What Inputs are Required	32
What measurements?	32
What makes a measurement useful?	32
How to take measurements?	33
Direct *vs* indirect measures	35
Measuring Locations	37
What animals to measure?	37
What other data should be collected?	37

Which identification system to use?	39
What added labour requirements are needed?	39
What management system	39
Managing Stock to Maximise Effectiveness of Recording	40
Planning the Whole Performance Recording Program	40
Evolution of the program	41
Further Items	41
Recording is not a spare time job!	41
GIGO operates!	41
Artificial breeding and performance recording	41
Graduating from within- to across-herd/flock recording	42
How many EBVs?	42
Dos and Don'ts in Performance Recording	43
The don'ts	43
The dos - Approach	44
The dos - Detail	44

PART II: GENETIC EVALUATION

4 Principles of Estimated Breeding Values   47

Brian Kinghorn

Heritability and EBV or EPD	47
Using Estimated Breeding Values	50
An Example - Yearling Weight in Beef Cattle	51
Dollar EBVs	52
Use of Dollar EBVs	52
Accuracy of EBVs	55

5 The Alternative Evaluation Procedures   57

Markus Schneeberger

Historical Development of Genetic Evaluation Procedures	57
Best Linear Unbiased Prediction (BLUP)	58
Statistical method	59
Classification of models	60
The Numerator relationship matrix	62
Obtaining the solutions	63
Accuracy of Estimated Breeding Values	64
The BLUP Animal Model	65
Genetic Evaluation for Categorical Traits	67

6	**Within-herd *versus* Across-herd Evaluation**	**71**
	Keith Hammond	
	Across-Herd or Flock Comparisons are Important	71
	Across-Herd or Flock Procedures	72
	Central Testing	73
	Sire Referencing	74
	Across-Herd or Flock Analyses	75
	Across-Herd or Flock Evaluation for Meat Yield and Quality	76
7	**Genetic Evaluation of Beef Cattle**	**77**
	Alex McDonald	
	Introduction	77
	Changing Attitude of Breed Associations	78
	Growth Traits	79
	Reproduction Traits	81
	Scrotal Size and Days to Calving	81
	Gestation Length	82
	Carcase Traits	83
	The Future	84
	Genetic Evaluation of Calving Ease for Australian Beef Cattle	84
8	**Genetic Evaluation in Wool Sheep**	**85**
	Michael Carrick	
	Introduction	85
	Within Flock Evaluation	85
	Current Woolplan	85
	The Future - Genetic Evaluation using BLUP	86
	Across Flock Evaluation	87
	Advantages and Disadvantages of Central and On-Farm Tests	88
9	**Genetic Evaluation in Meat Sheep**	**89**
	Robert Banks	
	Evaluation Structure in LAMPLAN	89
	Statistical Models for Genetic Prediction	90
	Selection Indices	91
	Across-flock Genetic Evaluations	92
	Delivering Genetic Evaluations through LAMBPLAN	92
	Summary	93

10 Genetic Evaluation in the Dairy Industry — 95
Michael Goddard

Calculation of EBVs for Milk Production Traits	95
Comparison Across Breeds, Herds and Age-Groups	96
Weighting and Standardisation of Yields	96
The Base	96
Overseas Bulls	97
Reliability	97
Interpretation of EBVs	98
Use of EBVs	98
Traits other than Production	99
Australian Dairy Herd Improvement Scheme (ADHIS)	100
Extension Problems	100
Why didn't my cow Flossy get an EBV?	100
Cow 1379 produced 200kg of fat last year while 1721 produced only 150kg, yet 1721 got the higher EBV. How can this be?	100
Why did the EBV of bull ABCD drop 10kg between 1990 and 1991?	101

11 Genetic Evaluation in the Pig Industry — 103
Tom Long

Introduction	103
Traits of Economic Importance	104
On-Farm Testing	105
Central Testing	105
Methods of Evaluation	106
Visual appraisal	106
Single trait selection	106
Selection index	107
BLUP and PIGBLUP	108
Fully Integrated Genetic Evaluation System	109
Guide for Consultants	110

12 Across-Breed Genetic Evaluation — 111
Andrew Swan

Across- *versus* Within-Breed Evaluation	111
Understanding Crossbreeding Effects	111
Simple Across-Breed Genetic Evaluations	114
More Complex Evaluation Procedures	115
Industry Application of Across-Breed Evaluation	117
Guide to Consultants	119

PART III: BREEDING OBJECTIVES

13 Introducing Economics to Modern Animal Breeding — 121
Stephen Barwick

Introducing the Breeding Objective — 121
Some Important Distinctions — 121
    Selection for More than One Trait — 122
Some Background to the Derivation of Selection Indices — 123
    A single-trait objective — 123
    The multi-trait objective — 125
Selection Index Examples — 126
    Example 1. One trait in the objective, two selection criteria available — 126
    Example 2. Two traits in the objective, two selection criteria available — 128
Selection for Compound Traits — 130
Using Economics to Formulate the Breeding Objective — 131
    Definition of an economic value — 131
    Describing the relativity of economic values — 132
    The measurement basis for economic values — 132
    Steps in defining the breeding objective — 133
Issues in the Derivation of Trait Economic Values — 134
    The level of detail attempted — 134
    Whose economics to consider? — 135
    Estimation methods — 135
    Effects of differing perspectives and management and marketing constraints — 135
    Non-linearity of economic values — 136
    The need for optimal management — 136
    Accounting for feed costs — 136
    Accounting for competitive position — 136
    Discounting — 137
Other Ways of Selecting for More than One Trait — 137
Guide for Consultants — 138

14 Breeding Objectives for Beef Cattle — 141
Stephen Barwick and Willi Fuchs

The Need — 141
B-OBJECT: A Breeding Objective and Selection Index Package — 141
    Basis of the procedure — 141
    Formulating the breeding objective — 142
    Identification of the breeding, production and marketing system — 142
    Identification of sources of returns and costs in commercial herds — 143
    Determination of the biological traits influencing returns and costs — 144
    Derivation of the economic value of each trait — 145
    Deriving the Selection Index — 146

	B-OBJECT Applications	148
	Who will use B-OBJECT?	148
	Further reasons for using B-OBJECT	148
	B-OBJECT Examples	149
	Guide for Consultants	152
	Steps in Using B-OBJECT	152

15 Breeding Objectives in Wool Sheep   155

Michael Carrick

Direction of Genetic Change is Important	155
Which Approach to Use	155
The Formal Approach Aims to Maximise Profit Increase	156
An Example	156
Integrating Economic Value and Genetics	157
Heritability - How Easy to Move Traits	157
Genetic Correlation - How One Trait Tends to Move Another	157
Variability - How much Room to Move Traits	158
Finalising the Selection Index	158
The Data Needed from the Breeder	160

16 Breeding Objectives in Meat Sheep   169

Robert Banks

Introduction	169
Industry Structure in Australia	169
Industry Use of Genetics	171
LAMBPLAN and Breeding Objectives in Meat Sheep	171
Approaches to Breeding Objectives in other Countries	174
Conclusions	175

17 Breeding Objectives for Dairy Cattle   177

Michael Goddard

Introduction	177
Economic Weights	177
Selection Criteria in Australia	180
Value for Money in Semen	181

18	Breeding Objectives in the Pig Industry	183

Tom Long

Direction and the Breeding Business	183
Defining the Goal	183
Traits Contributing to the Breeding Objective	184
$Index	186
Example - Terminal Sire Line vs Maternal Line	188
Guide for Consultants	191

PART IV: DESIGN OF BREEDING PROGRAMS

19	Principles of Genetic Progress	193

Brian Kinghorn

Manipulating Genetic Differences	193
Simple Selection Theory	193
Selection Intensity	197
Selection Based on other Sources of Information	200
Genetic Progress in Open Nucleus Schemes	201

20	Design of Straight-Breeding Programs - Common Problems	205

Keith Hammond

What are the Available Options for a Selection Program?	205
How Far Ahead to Plan?	207
Key Points in the Design of Selection Programs	207
Definition	207
Herd size	207
Selection	207
Generation turnover	208
Inbreeding depression	208
Mating	208
Records	208
Integrating the Selection Program and Management System	209
Common Problems in Selection Programs	209
Labour Requirements and Design of Selection Programs	210
Where are the Payoffs?	213
Single Genes in the Breeding Program	213
The breeding program for single gene traits	214
Record keeping	215

21	**Maximising Improvement with AI, MOET and Cloning**	217
	Brian Kinghorn	
	Introduction	217
	Improvements in Genetic Gain with AI and MOET	217
	Increased selection intensity	217
	MOET: Increased selection intensity and more information for Estimating Breeding Value	217
	Improvements Due to Cloning	221
	Use of AI and MOET in Open Nucleus Schemes	222
22	**Design of Crossbreeding Programs**	227
	Andrew Swan	
	Straight *versus* Crossbreeding	227
	The Benefits of Crossbreeding	227
	Problems Associated with Crossbreeding	228
	Crossbreeding Systems	229
	A Dynamic Approach to Establishing Crossbreeding Programs	231
	Practical Application of Crossbreeding in Livestock Industries	232
	Guide for Consultants	235

PART V: THE BREEDING BUSINESS

23	**Other Economic Considerations in Animal Breeding**	237
	Keith Hammond	
	Background	237
	The Seedstock Producer	237
	The Commercial Producer	239
	The Breeding has Been Done for Today!	239
24	**Management of Risk**	241
	Markus Schneeberger	
	The Risk About Selection of Replacements	241
	A Portfolio of Sire Usage	242

Index   247

❊ ❊ ❊ ❊ ❊ ❊ ❊

PART I: Modern Animal Breeding

Chapter 1

The New Era in Genetic Improvement of Livestock

Keith Hammond

The Fundamentals

Fundamental considerations to profitable livestock production are:

- The market to be supplied.
- Biological diversity.
- Productivity = Output per unit Input.
- Product quality.

The livestock producer will endeavour to maintain and even improve profitability by:

- Winning market share - Which are the best markets to supply in terms of amount and reliability of economic return?
- Increasing $ return per $ spent - This "terms-of-trade" as it is known, can be increased by increasing productivity. This means increasing output per unit of input and/or changing product quality when quality characteristics are included in the market signals.

Winning markets and increasing terms-of-trade can be achieved by:

- Utilising better equipment and better processes - Each link in the chain of operations from conception of the next generation of potential seedstock to the delivery of end product to the ultimate consumer.
- Obtaining and utilising better information - What to produce? How to produce it? How to market it? When to market it? Better information including both basic data and processed results, clearer signals and better decisions throughout the production chain may contribute to both short-term and longer-term profitability gains.
- Genetic change - Can also be an important means to win market share and increase terms-of-trade in a herd, flock or industry. This change may take the form of change in breeds used, change in crosses, and/or by changing sire and dam lines.

What is Animal Breeding?

Animal Breeding in Agriculture today and in the future is concerned with **manipulation of biological differences between animals over time** using approaches aimed at maximising profitability in both the shorter-term and the longer-term. Some of these biological differences are **genetic** or inherited, whilst others are **non-genetic** or environmentally induced, including some growth cycle characteristics, e.g. age is not inherited but has an impact on some economic quantity and quality traits. The genetic differences can be considered to occur at three levels, viz.

- Between breeds - Breeds differ in their average genetic composition often for many traits.
- Between crosses - Crosses of particular breeds may differ in up to three important respects: In their average genetic composition. Secondly, where the breeds contributing to the cross are substantially unrelated the genetic components for each of the range of traits we observe may interact in the cross to produce a multiplier effect known as **hybrid vigour** or heterosis. Of course the effect for a particular cross and trait may be negative or positive in terms of what is desired. Finally, for traits in crossbred progeny that result from the combined action of both the genes of parent and offspring, reversal of two breeds between sire and dam can lead to different end results in the progeny, e.g. weaning weight is the end product of both the calf's direct genetics and the dam's genetics for maternal capability, hence weaning weight of the crossbred calf may differ depending upon the breed of dam used.
- Between animals within breeds and within crosses - For most traits there exists very substantial differences in genetic merit among all the potential sires and all the potential dams of the next generation. This, together with the basic biological laws of inheritance, means there will also be very substantial differences amongst all the progeny of these parents.

The biological differences between breeds, crosses and animals which we are interested in manipulating are those associated with productivity and product quality. We may measure productivity and product quality **directly** by measuring traits which are directly associated with a cost, e.g. feed consumed, or with a return, e.g. carcase weight or fibre diameter; or **indirectly**, by measuring traits which do not result directly in a cost or a return but which are associated with one or more of the direct traits and are easier and cheaper to measure on younger animals, such as growth rate, scrotal size or even in the future molecular genetic marker traits.

In addition, Animal Breeding must be concerned with strategies which guarantee the producer longer-term access to genetic diversity, to enable profitability to be sustained. Of course, a range of strategies exist to do this. For example, breeding operations can be designed to ensure that a large **gene pool** persists, i.e. genetic diversity is maintained within the breeding population, or the producer can rely on introducing further genetic variation in the future from elsewhere, e.g. from elsewhere within the country, from different countries, from different sire and/or dam lines, from different breeds and perhaps in the distant future even from different species.

Basic Breeding Biology

The laws in inheritance which are central to modern animal breeding operations are straightforward but often still poorly accepted. Yet modern breeding technology and success in maximising genetic change in a herd or flock is heavily dependent upon just two of these principles, viz.

- Each offspring receives **a sample half** of its genetic makeup from its sire, and **a sample half** from its dam, and
- **Inheritance is imperfect** for the vast majority of traits associated with productivity and product quality.

The universal operation of these laws in livestock populations means that:
- Each offspring is a new genetic combination.
- Many characteristics change from parent to offspring.
- What you see isn't necessarily what you get!
- The offspring from two parents **are spread** about the average genetic merit of both, for all imperfectly inherited characteristics.
- Genetic change in the offspring will increase with increased accuracy of selection and with increased intensity of selection, for each trait of interest.

There are further general principles, for example, the primary mode of this inheritance is additive in action and the best expectation for the merit of the offspring is the parental average. There are also additions to the general principles which may need to be considered in particular situations, e.g. the importance of the multiplier effect of hybrid vigour for some traits in crosses of some breeds. However, thorough acceptance of the above-mentioned primary laws is universally important.

Does Animal Breeding Work?

The short answer is: Amazingly well, **if** the principles are followed!

Examples which may be used to demonstrate the nature of inheritance, and the importance of breeding are:

- The existing large differences, which are maintained over generations, between our domesticated livestock species, for a whole range of traits. Why do these differences exist?
- The large differences that exist between the range of breeds within each species. Why is this so?

- The large spread in estimates of genetic merit (Estimated Breeding Values or Predicted Differences) that exist in each successive crop of offspring, as shown by the use of modern genetic evaluation e.g. examine a table of percentile bands for trait EBVs for animals born in a particular year from a sire summary for a breed. Why does this large spread occur?
- The association of parent and progeny genetic evaluations. Why is this not perfect?
- The stability over time of genetic evaluations for particular animals. Why are female evaluations and those of younger animals generally less stable?
- The realised genetic changes that have been achieved and are now well documented in many industries and breeds for a whole range of traits in many countries. Why are these genetic changes greater than expected between some years (offspring crops) and less in others?
- The comparison of these observed genetic changes with those predicted for the breeding operation at the outset. Over numbers of generations these comparisons often show some disagreement. Once the reasons for this are understood, and almost universally the reasons relate to deficiencies in applying the general principles, they can be used to improve the effectiveness of the breeding operations and as educational material.

Developments Relevant to Modern Animal Breeding

Developments which have enabled increases in the effectiveness of Animal Breeding programs are summarised in Figure 1.1.

Modern Animal Breeding programs, planned on established genetic principles, are very recent indeed, the first being initiated only 40 to 50 years ago, depending on the species. There has only been time for 10 to 15 generations of cattle selection and 40 to 50 generations of selection for egg number, growth rate and feed efficiency in chickens.

However, this recent period, when established (if basic) genetic theory was first applied to animal production, was preceded by several thousand years of domestication during which both natural and human directed artificial selection occurred in isolated groups of animals, resulting in large numbers of 'old world' breeds that differ in many characteristics.

The Englishman Robert Bakewell provided much of the impetus for further breed development and for the formation of herdbooks and pedigree selection in the early 1800s. The formal principles of single gene Genetics were not initiated until the research and publications by the Austrian monk Gregor Mendel in the 1850s. However these publications then remained 'undiscovered' until the early 1900s.

The first organised performance recording was initiated in the 1880s. At about the same time experimental programs were commenced to select for oil content in corn populations in the USA. This is mentioned because it is an outstanding example of the large amount of quantitative genetic variation that can exist for a trait. Rapid and linear response to selection continued in these lines for many decades.

Since 1950 a spate of events important to increasing the efficiency and effectiveness of Animal Breeding programs have occurred; and there is no let up!

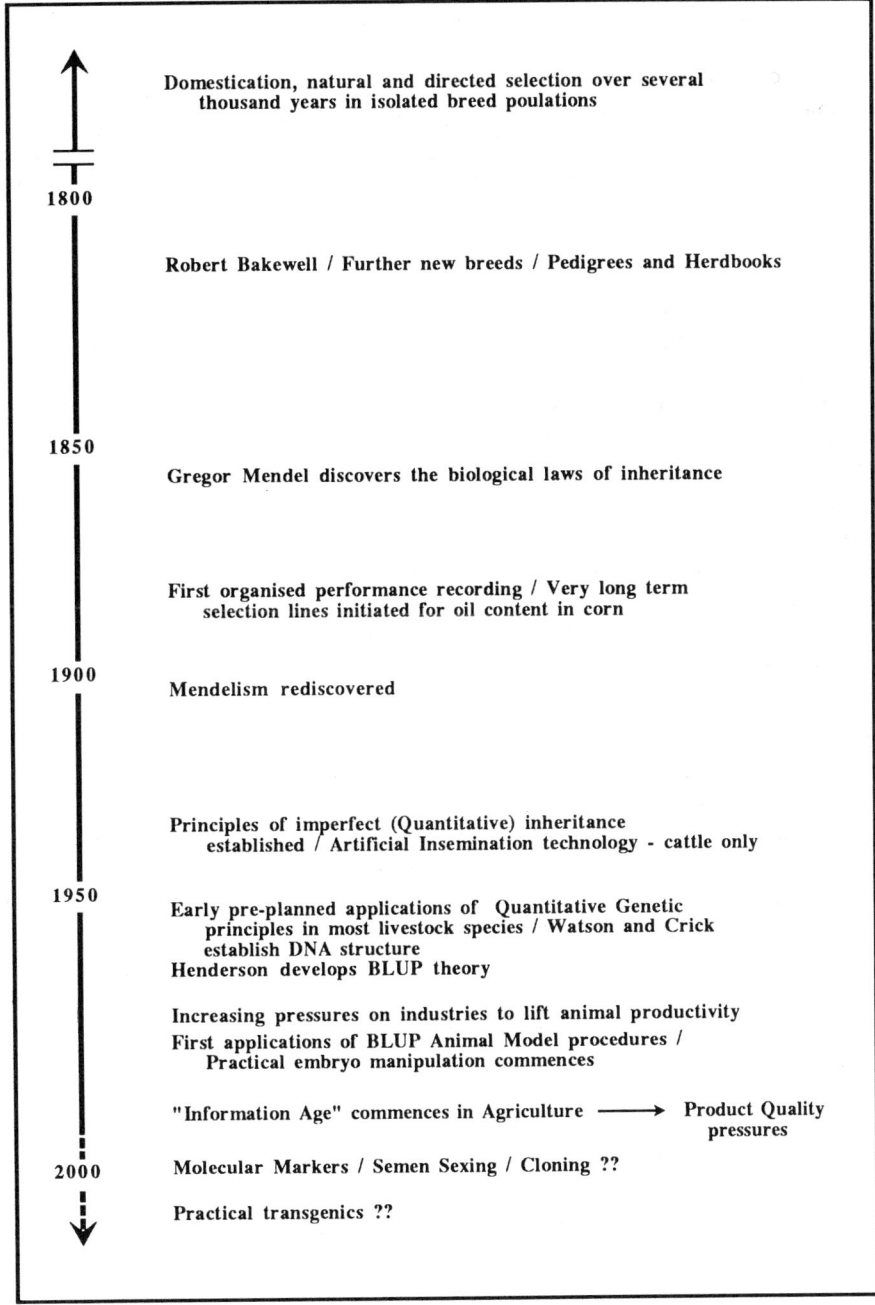

Fig. 1.1 Some important events in the development of modern Animal Breeding

The structure of DNA was established by Watson and Crick in the early 50s, leading to many subsequent jumps in our understanding of the biology and to important new procedures for studying it and, eventually, for manipulating it in the field.

Also in the early 50s the late Professor Charles Henderson from Cornell University commenced the development of the mixed model statistical theory, the linear form of which is known as Best Linear Unbiased Prediction or BLUP. This "gave genetic theory a natural structure and efficient properties; indeed the basic premises of BLUP are so simple and its applications are so general as to make it a most elegant scientific structure, not just a useful tool" (Hill, 1987). Experience to date emphasises that the real significance of BLUP in the field is not its ability to produce generally modest increases in the accuracy of selection compared with other possible approaches, or even to enable producers to receive estimated genetic trends for each trait of interest in addition to the Estimated Breeding Values (EBVs) they obtain for every animal, but it can generate a high level of end-user comfort, and consequently acceptance of the technology. This is because it is able to cope with a range of conditions which were not properly accommodated by previous breeding technology.

In the 60s and 70s, depending on the particular livestock industry and country economy and markets supplied, pressures intensified to increase animal productivity rather than simply increase production. The first applications of the more advanced and useful BLUP models, the so-called animal models commenced, embryo manipulation became practical in a limited way. It is interesting that although artificial insemination for cattle was practical under some circumstances during the late 1940s its large scale use for sheep and pigs has only become a reality during the 1980s. AI is likely to remain the most powerful of the current and possible artificial breeding technologies.

During the 80s the 'Information Age' commenced to take hold in agriculture, as computing and communications technologies became practical. This is a particularly important development for facilitating better Animal Breeding because of its potential to both help generate and clarify market signals, from consumer of end product through retailer, wholesaler, processor, buyer, possibly several levels of commercial producer and finisher, to seedstock multiplier and eventually to the initial seedstock producer.

In the near future, a range of molecular genetic markers will be developed for use in increasing the power of selection, particularly for those traits which are costly to measure, not measurable in young animals or measurable only after slaughter or in one sex. In addition, the introduction of semen sexing will promote changes in industry structure although it will only offer small increases in genetic change. Likewise, cloning of embryos, which we can expect to become practical in the mid-90s, will offer important possibilities for changes to livestock industry structure and also for evaluating carcase and meat quality traits which cannot yet be measured on the live animal, so preventing the collection of direct carcase information on all potential breeding stock. These developments will be supplemented by engineering of new and cost-effective equipment for use in simpler and more direct measuring of traits prior to slaughter and earlier in life.

Current developments in molecular biology also offer the potential for a range of other technologies, some of which will impact on Animal Breeding, e.g. a 'litmus paper' test for use in the field to detect carriers of harmful recessive genes and to confirm the sire and dam of animals produced under multiple sire mating, and eventually transgenic animals themselves will play an important role in at least some of our livestock industries.

However, the correct perspective for the future of Animal Breeding is one where the coming molecular technologies complement the 'Modern Animal Breeding Approach' which is dealt with throughout this text. Whilst ever food and fibre products are produced from animals the principles outlined in this text will be relevant!

Information in Animal Breeding

Achieving genetic change in animal populations involves:

- Collecting, processing, and interpreting information to make decisions.
- Acting on these decisions to eventually reap rewards from the resulting genetic change and to generate feedback to improve the efficiency and effectiveness of this **breeding information system**.

The flow of information in this system is diagrammed in Figure 1.2. In developing a breeding program primary considerations associated with this flow of information, decisions and actions are:

- What information, decisions and actions are needed?
- How are they to be organised cost effectively?

Of course, information systems are not unique to Animal Breeding programs. However, once the source of genetic material to be used in a breeding operation is decided upon then the success of the program as a whole is vitally dependent upon the type and quality of information collected, the way it is processed and interpreted in arriving at the decisions, and the efficiency of the actions necessary to implement the decisions.

To start with superior genetic material is an advantage, but it will then be the design and efficiency of implementing the information system which determines the speed of genetic change in the breeding program. This is often poorly appreciated. It is also the reason why modern breeding programs will continue to be increasingly focused on regional, national and eventually even international breeding information systems. Examples are BREEDPLAN International, WOOLPLAN, LAMBPLAN and the Australian Dairy Herd Improvement Scheme.

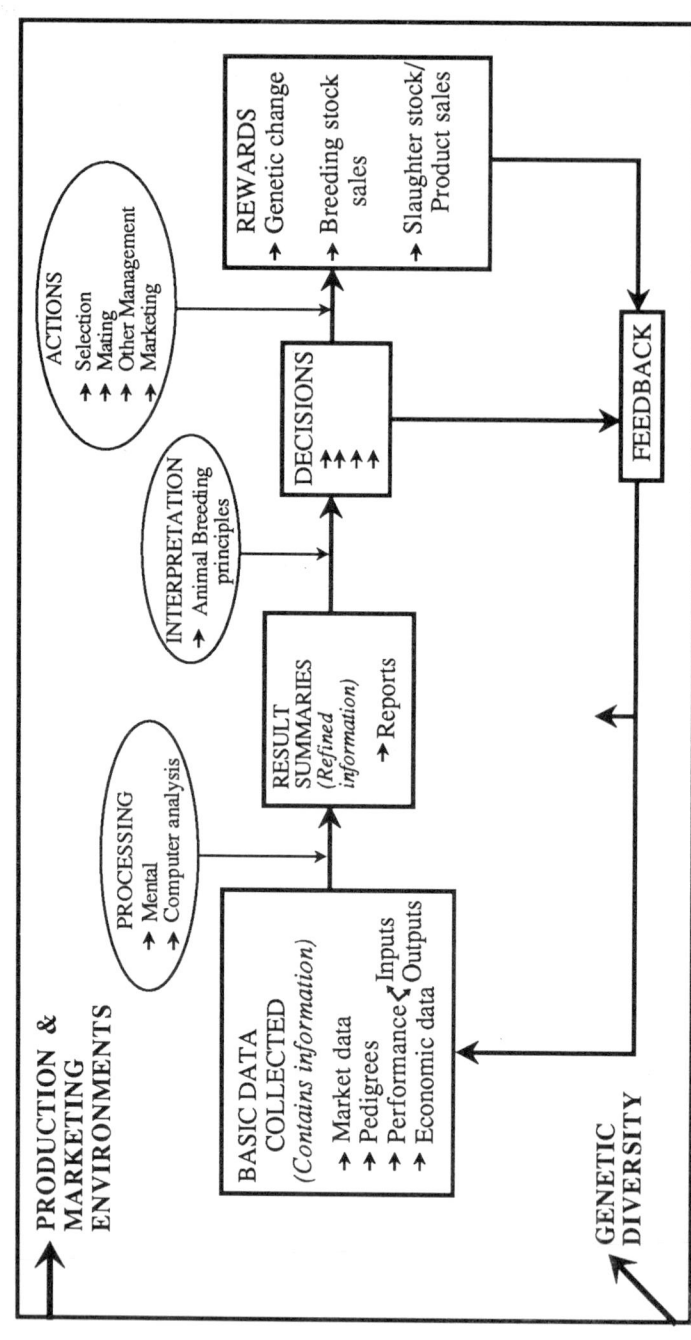

Fig. 1.2 Information flow in breeding programs

It is becoming increasingly recognised that pedigree information, or performance information is not sufficient alone to operate a breeding program. A range of marketing and economic data is also needed, both at the outset and during the breeding process. Likewise, with increasing recognition of the importance of productivity gains to help overcome the deteriorating terms-of-trade of many animal industries, and to use more effectively our finite world resources, the inputs animals require are being increasingly and correctly considered as components of performance, in addition to the measurement and recording of their outputs. Of course, not every variable needs to be measured by using a piece of equipment. Some data can be collected and processed using mental power alone and of course, the physical resources required in actually inspecting each animal; whilst in other cases data may be collected by subjectively scoring traits but still using computer analysis to produce the reports which are subsequently used to make the necessary decisions, e.g. this is often done when breeding for ease of calving. Important here also are the guidelines for designing such scoring systems to be effective. These will be considered in a later chapter.

The New Era in Animal Breeding

During the past decade or so a number of important technological developments have occurred which are changing the way Animal Breeding programs are operated and, not surprisingly, increasing their effectiveness. Some of these developments have actually resulted in modest increases in the genetic change predicted from theory, whilst most have resulted in improved capability to achieve the theoretical expectations in practice. Brief reference to these was made in Figure 1.1, whilst Figure 1.3 summarises the most significant areas of development, together with the major outcomes in the field.

We shall not further discuss here these developments and outcomes, for they will be treated at various points throughout this text. However, the changes in Animal Breeding operations they are enabling, at the herd or flock, region, national and international levels, are such that we can unequivocally say they are heralding a new era in the genetic improvement of our major livestock populations.

Taking on New Technology

These new developments will offer many challenges to breeding organisations, seedstock producers and even buyers of seedstock. Of course, no breed or seedstock producer needs to be involved, but they will need to compete with those that do!

With the opportunities and increased potential for genetic change in practice offered by some of these new technologies also comes the higher risk from misuse. The challenge is to do it right!

DEVELOPMENT AREA	OUTCOMES
• ELECTRONIC / COMMUNICATIONS ENGINEERING	→ Increased consumer pressures for product description and specification. → Better transfer of market signals throughout the production chain. → Better measures of productivity and product quality. → Automated identification, feeding and management of animals to facilitate performance recording / reduce labour costs. → Better processing of data.
• STATISTICAL ANALYSIS PROCEDURES	→ Better decision aids, e.g. EBVs and Genetic Trends from BLUP analyses.
• REPRODUCTION TECHNOLOGIES	→ Easier, quicker, cheaper transport of genetic material within and between countries. → Increased multiplication of (superior?) genotypes. → New opportunities to change industry breeding and production structures. → Increases potential for genetic change from: better linkage of herds and flocks for genetic evaluation, increased accuracy and intensity of selection, reduced generation intervals and reduced inbreeding and risk if managed correctly. → Necessary tools in the use of molecular genetics.

• MOLECULAR GENETICS	→ Many possibilties. In the near term:
	  → Increased accuracy of selection utilising molecular markers.
	  → Simple field teasts for detecting animals carrying deleterious alleles.
	  → Simple field tests for parentage confirmation.
	  → Improved understanding of the biology of production.

Fig. 1.3 Primary technological developments associated with the new era of Animal Breeding

Often breed associations feel threatened by impinging new technologies and regulate in an effort to control the spread of these technologies. There are already some prominent examples of this with the Artificial Insemination (AI) and Embryo Transfer (ET) technologies. Again, care is needed as regulations to limit or otherwise control the use of new technologies also have the potential to place the breed, and of course its members, at a comparative disadvantage to other breeds that allow the market to operate.

Care is also needed in utilising these new developments, particularly because our understanding of the biology of production in most species still involves some substantial 'black holes'. Often the level of ignorance tends to be overlooked. Of course, the new technologies will often also be used in the research environment to help overcome this ignorance.

Product Pricing Systems and Genetic Improvement

Animal Breeding programs are directed at improving profitability and can generally be considered as longer term investments. Hence, the price signals being returned from the market place are particularly important to the design and effectiveness of breeding operations. In all industries the Information Age is forcing pricing systems to become more objective, and is certainly clarifying price signals. Nevertheless, there remains much 'noise' in most pricing systems associated with animal production. The pricing systems often still used for milk are good examples. Government regulation is used to control supply and maintain continuity of supply; and commonly the pricing system does not properly recognise all four major components of milk, viz. butterfat, protein, lactose and the carrier water. Yet the energy input costs for these major components are very different. Approximately 70:36:25:0 megajoules of feed energy is required to produce a kilogram of butterfat, protein, lactose and carrier, respectively. In addition, the processing costs for these four components commonly vary greatly and modern processing technology enables each to be either added or taken from the milk. Finally, the value in the market place of these four milk components varies substantially. The milk pricing system which is fairest to all and provides greatest opportunity for increasing productivity at each link in the production chain would correctly accommodate each of these factors.

An ideal pricing system then will:
- Recognise the range of consumer requirements.
- Maximise opportunities for productivity gain at the farm and processing levels.
- Contain separate signals for each of the major components, based on their market value.
- Relate to the total raw product.
- Recognise quality differences including contaminants.
- Recognise seasonality of production costs.
- Recognise variable transport and processing costs.

Overview

This chapter has defined Animal Breeding and placed the range of formal disciplines involved in perspective with its fundamental imperative to achieve biological gains in short-term and longer-term profitability. The chapter has also provided an historical overview which highlights the current rapidly increasing potential for achieving and disseminating genetic change in many of our livestock industries. It has highlighted the importance of information and of the decision processes to the total Animal Breeding operation. In the next chapter we identify these primary decision processes and define a simple approach to Animal Breeding which accommodates all these primary processes.

Reference

Hill WG (1987) Increasing the rate of genetic gain in animal production. The Sixth AS Nivison Memorial Address, University of New England, Armidale. 19 February, pp 20

Chapter 2

The Modern Breeding Approach

Keith Hammond

Introduction

In Chapter 1 we defined Animal Breeding as the manipulation of biological differences between animals over time to maximise profitability. We discussed the importance of information flow to the total breeding operation (also see Figure 1.2), in:

- Processing the range of basic data which has been collected;
- Interpreting the results of this processing, in arriving at the necessary decisions; and
- Acting on these decisions to reap rewards from: achieving genetic change, selling breeding stock, and marketing more product of better quality.

A logical approach to Animal Breeding then is to ask: What decisions are required to maximise profitability over time? Having identified these decisions we are then in a position to backtrack through the information flow to determine precisely what basic data is necessary for the breeding operation.

In this chapter, we shall first identify the key decision areas in Animal Breeding before forming out of these three processes the primary operational components of our 'Modern Breeding Approach' to achieving genetic change.

The **set of nine key decision areas** together with the **three primary components** offer a simple, logical and comprehensive conceptual and practical approach to Animal Breeding. This can be used when considering the development of breeding operations, whilst generally leaving in the background the very substantial body of genetic theory which now exists and which often confuses people in the field.

In addition, **checklists** can be developed based on these key decision areas and primary components, for use in current programs. Again, these lists can be superimposed on the theory to reduce its profile in practical situations.

What are the Key Decision Areas in Animal Breeding?

The nine key decision areas which confront producers when considering their breeding operations are listed in Figure 2.1. A brief summary of each area is provided here. Considerable detail on each is given throughout the remaining chapters of this text with, for example, several chapters being devoted to the key decision: What is Improvement?

1.	**TO BREED OR BUY REPLACEMENT STOCK ?**	○ To produce/market seedstock or to market only unfinished/finished commercial product.
2.	**WHICH BREEDING ENTERPRISE TO PURSUE ?**	○ For example, straightbreeding and/or production of crossbred sires and dams.
3.	**WHAT IS IMPROVEMENT ?**	○ In terms of maximising profit? (The breeding objective for the herd/flock).
4.	**WHAT RECORDING SYSTEM ?**	○ A complex of questions/decisions.
5.	**WHAT TO CULL ?**	○ To maximise gains from the current herd/flock.
6.	**WHAT TO SELECT ?**	○ To maximise genetic gains in progeny, grand-progeny, etc.
7.	**WHAT TO MATE ?**	○ Including age at first mating, mating structure and the use of artificial breeding technologies.
8.	**HOW EFFECTIVE IS THE WHOLE BREEDING PROGRAM ?**	○ In terms of maintaining cash flow, and maximising long-term return on investment.
9.	**HOW TO MERCHANDISE ?**	○ Both seedstock and commercial product.

Fig. 2.1 The 9 key decisions areas in Animal Breeding

Area 1: **To breed or to buy?** If the decision is to breed then the subsequent decision areas also apply. Because animals differ genetically, the buyer of seedstock who aims to breed stock for commercial product will also need to decide which animals to buy if profit is to be maximised. Even the secondary buyer of feeder/store stock can increase profit margins by deciding which lines, breeds, crosses, strains or progeny groups to buy, so capitalising on genotypic differences between lines, and which animals within the chosen lines, to utilise the phenotypic differences between the animals in the lines.

Area 2: **Which breeding enterprise?** For example, straightbreeding with Angus for young cattle production, or crossing Brahmans and Limousins to breed heavy steers, or a two breed rotational crossing of Hereford and Jersey to supply markets for crossbred females and feeder steers which marble well.

Area 3: **What is improvement?** To establish the breeding objective for the chosen enterprise.

Area 4: **Which recording system?** This comprises a complex of questions, viz. how and which animals to identify, which measurements, including visual scores, to make, what other data should be recorded, e.g. management groups, mating data, and how should all data be managed. Data management covers a range of operations from measurement, through checking, transferring, storing, retrieving and again editing data; and arranging for the regular analyses for timely presentation in forms which simplify the following regular decision processes.

Area 5: **What to cull?** To maximise profit from the current herd or flock and to maximise the phenotypic gain.

Area 6: **What to select?** To produce the next crop of potential replacement breeding stock some young males and females will be selected to be sires and dams because their EBVs are superior to some of the present sires and dams which will be culled, whilst other well-used sires may be culled to reduce inbreeding in future generations even though they may still number amongst the best males available. In smaller herds and flocks in particular it may also be necessary to decide whether to increase the number of male parents selected, to hedge against sampling and genetic drift causing less progress.

Area 7: **What to mate?** Decisions cover the age at first mating and sometimes mating frequency, e.g. once or twice per year establishing the actual mating pairs or groups and, increasingly, as the range of practical artificial breeding technology develops, the type of mating. These decisions will also influence the structure of the breeding program, so they will impact on the selection differential, the generation interval, the build-up of inbreeding and the chance of success and, via the assortment of pedigree information, the range in accuracy of future EBVs.

Area 8: **How effective is the breeding program?** Two decision processes are involved, concerning the **maintenance of cash flow** from the enterprise and **longer term progress and return on investment**. Quite often the impact of breeding and of alternative designs is evaluated by considering only longer term gains yet breeding operations can influence cash flow dramatically, their impact commonly differing between designs.

Area 9: **How to merchandise the results?** For the seller of breeding material in particular, this becomes an increasingly important consideration as competition intensifies and the costs of promotion increase.

These nine key decision areas are not independent. Some are closely associated, e.g. the culling, selection and mating decisions, the interplay of cash flow considerations with those concerning longer term gains, and merchandising should relate to the breeding objective; with these three sets of decisions also drawing from information on the effectiveness of the breeding program (essential feedback - see Figure 1.2). Further, in practice, all the above decision areas will need to be addressed repeatedly, initially and throughout the life of the breeding operation.

The Three Primary Components of Breeding Operations

Given the above decision areas and assuming desirable field practices, we can obtain three primary operational components of the breeding program. These, together with the linkages among them, are shown in Figure 2.2.

This division of breeding operations into three primary components offers a simple and practical conceptual approach for manipulating genotypic variation within and between populations of animals; although, to date it has only been applied to within-population selection programs. When utilised, the approach is capable of supplying to the industry, herd or flock all the key breeding decision aids necessary to reliably generate rapid gains in productivity and product quality, and in a ready and easy-to-use form.

The three primary components can cover all formal and practical aspects of breeding. However, the approach requires that cost-effective and accurate measurements exist which relate to all traits in the breeding objective, i.e. it assumes that an effective performance recording system is in place. Some industries still lack cost-effective and accurate measures of some of the important economic traits, e.g. feed intake and meat quality particularly on live animals, for seedstock producing herds seldom slaughter large numbers. Better direct and indirect measurements of reproduction, production and product, including low cost measures, are now been taken on the whole herd or flock early in life, and automated measures are now an increasingly actively area of research, development and application.

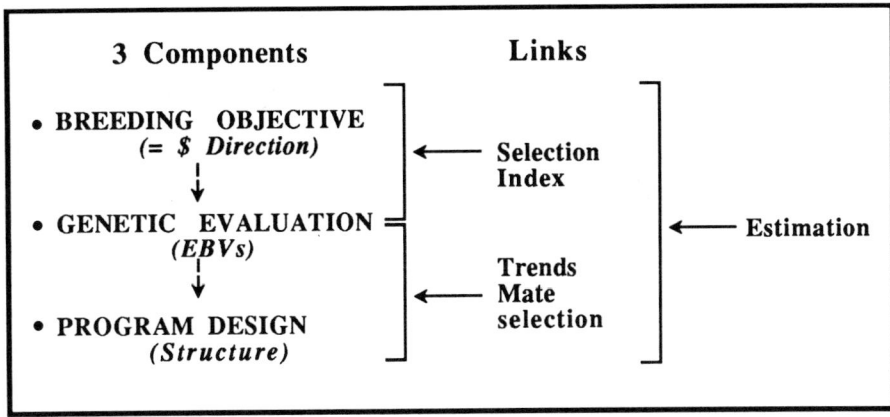

Fig. 2.2   The 3 primary components in the Modern Breeding Approach

We briefly comment here on each of the three primary components. Again, they are each dealt with in detail in one or more of the succeeding chapters.

Component 1:   **The breeding objective.** Establishes the direction to breed in economic terms: the '$ Direction'. This is often done intuitively by seedstock producers and buyers - no calculation, just by guess and "experience". It can also be done using formal calculation, and is already done this way in some cases. In the future it will be increasingly necessary to use a formal approach if the breeding operations are to maximise the exploitation of genetics. Computer packages for individual breeding operations to use in doing this have now been developed for some industries, e.g. the B-OBJECT package, and the $INDEX module in PIGBLUP, both developed by the Animal Genetics and Breeding Unit (AGBU) for use in the beef and pig industries, respectively.

Component 2:   **Genetic evaluation.** Provides the predictions of genetic merit the Estimated Breeding Value (EBVs), Australian Breeding Values (ABVs) or Estimated Progeny Differences (EPDs) for each animal for the measures and combinations of measures; and, if required, the estimate of risk (accuracy) associated with each EBV. Genetic evaluation enables animals to be ranked for individual measures and their components, e.g. the two EBVs computed from weaning weights for growth and milk, and on their overall economic merit for a particular breeding objective.

Component 3: **Breeding program design.** Establishes the optimum mating structure, including numbers of females per sire, and amount of selection, and the optimum period parents are used in the herd or flock, breed or sector of it.

Components 1. and 2. are linked by the **selection index,** that formally or mentally derived combination of measures - or separate EBVs - which has the maximum association with $ Direction. Components 2. and 3. are linked by the **genetic changes or trends** achieved in the breeding operation; and by a current development termed **mate selection.** Mate selection procedures will not simply rank individual animals on their estimated genetic merit, as EBVs do, but will identify the mating combinations which best fit the $ Direction, the breeding objective, established for the enterprise and, where necessary, will also make provision for inbreeding.

Finally, **estimation** of the parameters which are then repeatedly used in the analyses involved in applying each of the three components links all the components. Examples are the economic values, (co)variances, heritabilities and correlations, and breed and cross differences. Estimation then provides the formal 'backbone' to the total breeding operation.

Breeding information systems are now being developed to accommodate these three components and the analyses they involve. Sometimes this is done through a central bureau service, for example the genetic evaluation component known as BREEDPLAN International has been combined with a data management and processing module for breed societies. In other situations, e.g. with PIGBLUP for the Pig industry, B-OBJECT for the Beef industry and LAMBPLAN, the analytical software package has been designed to be portable on microcomputers and either used by field consultants or breeders themselves to undertake their own analyses for breeding objectives or for EBVs and trends. Hence, the strategy for implementing all components of a breeding information system can vary both between and within industries. Later chapters will deal further with these aspects.

The Extent of Genetic Differences

We have said that genetic change in productivity and product quality involves identifying and manipulating genetic differences between animals for the traits of interest. Figure 2.3 provides a simple demonstration of the extent of these differences in a large population of animals. It shows, for cattle for example, that the 1,300,000,000 (or 1.3×10^9) cattle in the world very likely all differ genetically. The same argument can be used for other species.

Let us expand a little on Figure 2.3 The number of genes carried by every cow, sheep, pig etc. is about 100,000 (or 10^5). Let us be very conservative and assume that for this particular cattle example only 100 of the 100,000 are segregating, i.e. more than one gene form (allele) exists in the population for each of these 100 genes, with the remaining 99,900 genes being fixed, with one allele only, in this population. The possible number of genetic types (genotypes) in the population is then 3^{100} or 10^{48}, assuming that there are only two alleles segregating at each of the 100 gene loci. Remember, with two alleles there are three possible genotypes, two homozygous types and one heterozygous genotype. 10^{48} is a very big figure indeed!

There are some assumptions in this calculation, but the answer is probably still conservative! The assumptions are:

- Each allele is transmitted from parent to offspring completely independently. However, the 10^5 genes in cattle are located on 30 pairs of chromosomes, so 'strings' of alleles for many genes tend to be transmitted as 30 blocks. The number of 'blocks' are much greater than 30 because crossing-over occurs at the formation of gametes with the 'broken' strings recombining across homologous pairs into a new chromosome string of alleles. This occurs with each of the 30 chromosome pairs at gamete formation. Hence, new combinations of each chromosome pass from parent to offspring, rather than one complete member of each pair of each parent's chromosomes.
- There are just two allelic forms for each segregating gene. Often there are more than two alleles of a gene segregating in a population, so producing more than three possible genotypes for each gene.
- Genes act independently. Some genes interact (Epistasis) to create further differences between animals.
- The expression of alleles is constant across all environments. Alleles can change in their relative level of expression between environments, producing further possibilities in terms of the observed final expression (phenotype).

The result is: Every animal in the world almost certainly differs genetically. The only exceptions are true identical twins. Hence, the pool of genetic differences available to manipulate in a species or breed is truly massive, **much larger than visually apparent** from a group of animals. We generally grossly underestimate the extent of these differences, because we tend to concentrate on a small number of characteristics of the outward appearance of animals.

Which Genetics is Important in Animal Breeding?

Genetics is a conglomerate science. **Population Genetics** involves the study of the behaviour of genetic variation in populations and of the forces which change gene frequencies, viz. mutation, migration, selection, and random drift of frequencies. The simplest form of Population Genetics is the **Simple** (or **Mendelian**) **Genetics** which deals with the inheritance of traits under the control of just one or two genes. This includes, for example, a range of inherited abnormalities, some simple forms of coat colour, particular milk proteins and a small number of other favourable genes of very large effect, e.g. the Booroola gene in sheep. Even the inheritance of some of these apparently simple characteristics is more complex than first appears. Whilst there may be just one or two genes of large effect segregating for a trait such as polledness or umbilical hernia, the final expression is often determined by the genetic background provided by a good number of other genes, so the inheritance is seen to be imperfect as it is with the quantitative traits.

What is the chance of all animals being different ?

- 1 Gene $\longrightarrow$ 3 Genetic Types = 3^1

- Each Animal $\longrightarrow$ 100,000 Genes = 10^5

- Say only 100 of these 10^5 genes have alternative forms segregating

- 100 Genes $\longrightarrow$ 3^{100} = 10^{48} Genetic Types

Compared with 10^9 cattle in the world !

SO: <u>ALL ANIMALS DIFFER</u>

What assumptions are in the above?

    1. *Each allele is transmitted independently from parent to offspring*
    2. *Only 2 alleles per gene*
    3. *Genes act independently*
    4. *Constant expression of alleles across all environments*

Fig. 2.3 A simple demonstration of the extent of genetic differences in a species

Quantitative Genetics comprises the other major component of Population Genetics, and of the operational science, Animal Breeding - see Figure 2.4. The vast majority of the traits of interest are quantitative (polygenetic). Their expression is determined by many genes, and they are incompletely inherited. Hence the need for the use of statistical distributions to help interpret their biology, and for the concepts of heritability; repeatability; genetic, environmental and phenotypic correlations; the concept of breeding value and its accuracy, or reliability or other measure of dispersion about the breeding value; and the concept of heterosis.

The 10^5 or so individual genes which form each genotype act in various ways when present in their heterozygous state, but for the quantitative traits we are primarily interested in their additive effects, with the exception of between-breed differences or variation, where we are commonly also interested in the non-additive (heterotic) effects in the crosses.

Because many of the biochemical processes behind animal physiology impact on the final expression of more than one productivity and/or product quality trait, and because genes are linked on chromosome strings, the genetic differences in each economic trait of interest are often not inherited independently of other traits. The superior genetic differences for two traits may be inherited together they have a positive genetic correlation, e.g. butterfat yield and protein yield in milk, or divergently they have a negative genetic correlation, e.g. wool yield and fibre diameter. Hence, both the heritability of and the genetic correlation between all pairs of traits are of particular importance in Animal Breeding. Figure 2.5 depicts these associations.

```
┌─────────────────────────────────────────────────┐
│                                                 │
│           WHAT'S IT ALL ABOUT ?                 │
│                                                 │
│              Animal Breeding                    │
│           (The Operational Science)             │
│                                                 │
│                ↗           ↖                    │
│  Simple(Mendelian)  ---->   Quantitative        │
│       Genetics         ↑      Genetics          │
│               ↖        |     ↗                  │
│                 ╲      |    ╱                   │
│                  Molecular                      │
│                   Genetics                      │
│                                                 │
└─────────────────────────────────────────────────┘
```

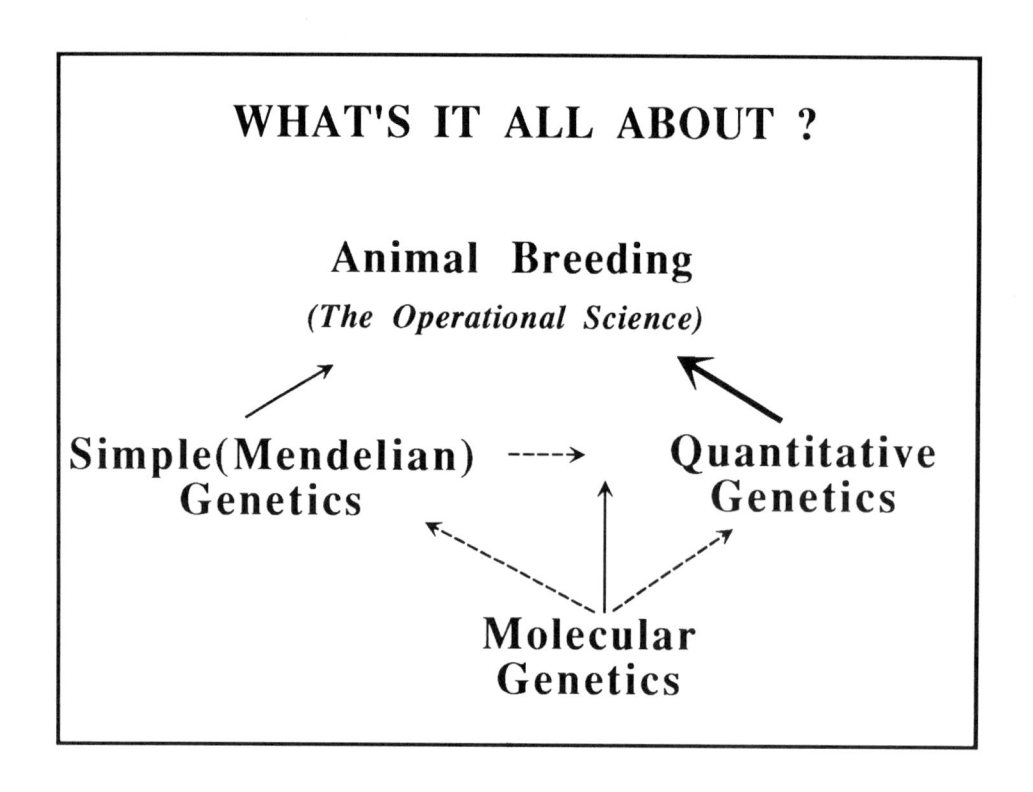

Fig. 2.4 Genetics and Animal Breeding - which differences are important?

Molecular Genetics involves the study, use and manipulation of genetic differences at the molecular level. Not surprisingly then there is great potential power and many possible applications for Molecular Genetics. Figure 1.3 listed just some of these. Molecular Genetics is still very immature, particularly for use in animals. Nevertheless, early practical applications are commencing in some species and these developments will rapidly gather momentum.

These then are the main areas of Genetics, although we often see the conglomerate further partitioned into disciplinary areas such as Physiological Genetics and Immunological Genetics; by species, such as Pig Genetics and Dairy Cattle Genetics; and even by professional area, such as Veterinary Genetics.

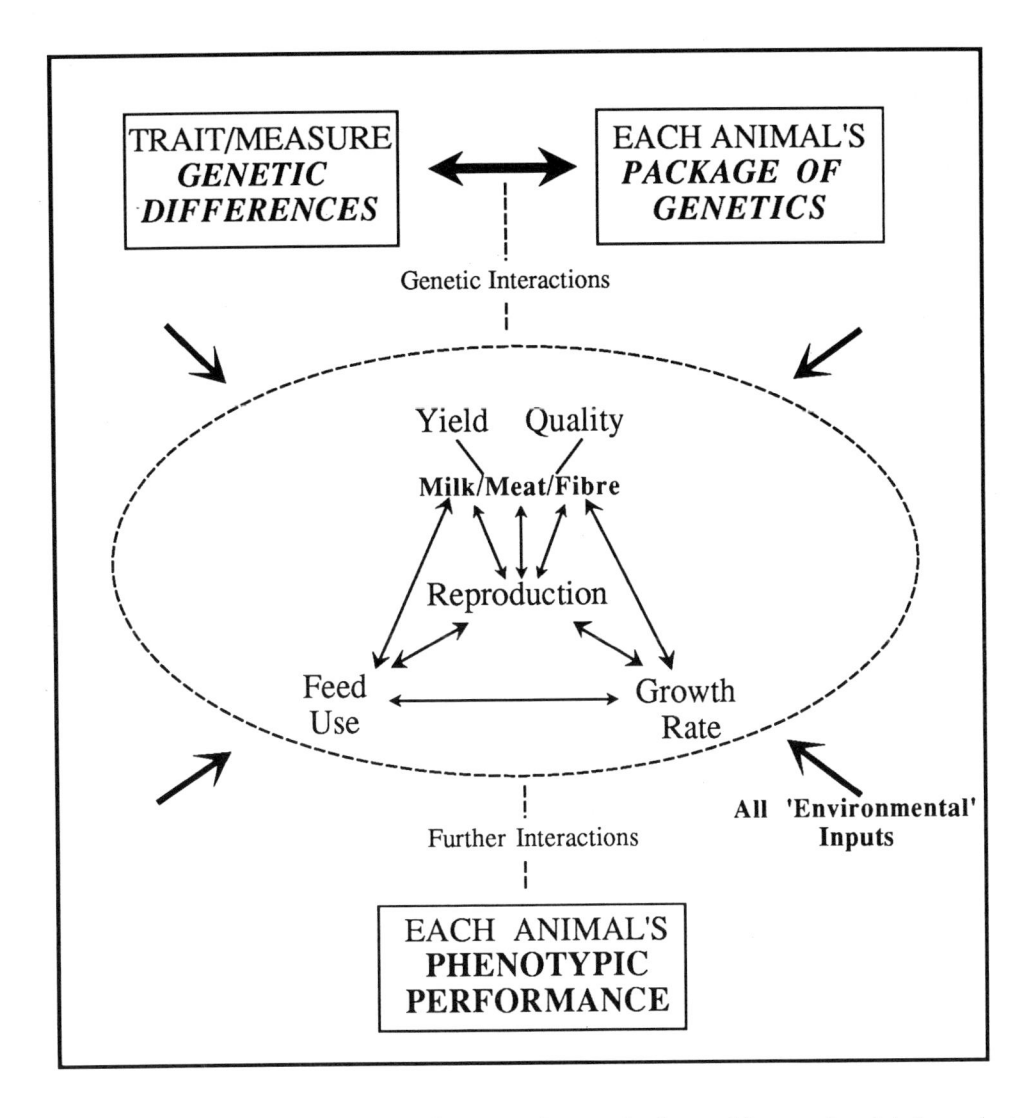

Fig. 2.5 The genetic differences for two traits may be favourably associated, independent, or antagonistic

Some Poorly Understood Concepts in Animal Breeding?

At **each** particular level of environmental inputs there will be genetic differences between breeds, between crosses, and between animals within breeds and crosses, for the productivity and product quality traits. This is depicted in Figure 2.6. The concept of improving management until animals have reached their genetic potential and then worrying about upgrading the genetics is false.

Another misunderstanding involves the idea of reducing the number of traits under selection in a breeding program to a bare minimum because as more are included then less genetic change in each is achieved. Firstly, all traits of economic importance should be included in the breeding objective, otherwise the $ Direction which is set to maximise profitability gain will be wrong! Secondly, the measurements to be included in the recording operations, for subsequent use in selecting replacement breeding stock, are those which can be recorded easily and cheaply in practice and which in combination maximise gains in the breeding direction, i.e. in profitability.

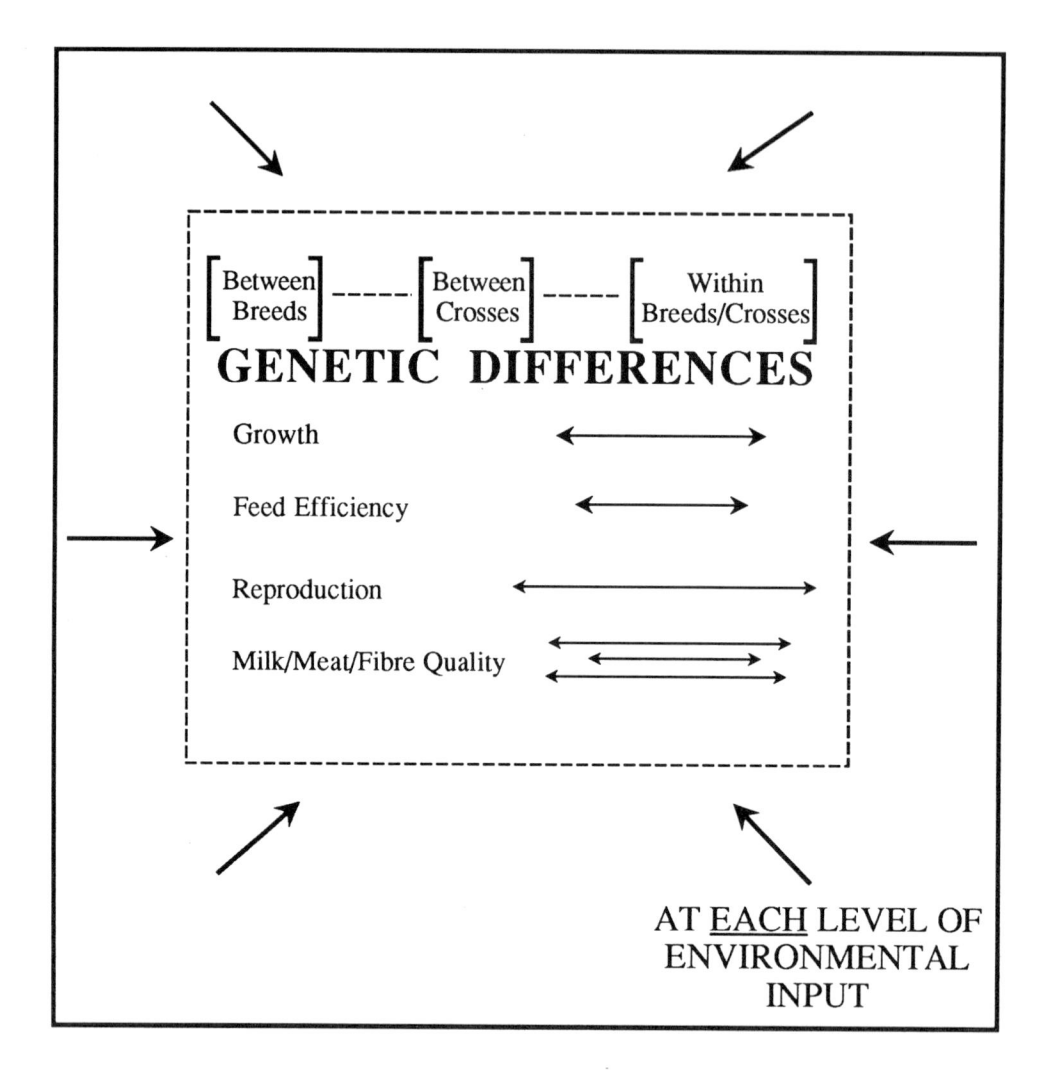

$$\begin{bmatrix} \text{Between} \\ \text{Breeds} \end{bmatrix} ----- \begin{bmatrix} \text{Between} \\ \text{Crosses} \end{bmatrix} ----- \begin{bmatrix} \text{Within} \\ \text{Breeds/Crosses} \end{bmatrix}$$

GENETIC DIFFERENCES

Growth

Feed Efficiency

Reproduction

Milk/Meat/Fibre Quality

AT <u>EACH</u> LEVEL OF ENVIRONMENTAL INPUT

Fig. 2.6 At each level of environmental inputs there will be genetic differences in productivity and product quality

A current popular misunderstanding is that Molecular Genetic procedures will be developed to the point where they will replace the current mainstream use of Quantitative Genetic principles in Animal Breeding operations. This will not occur whilst ever milk, meat and fibre are produced from animals! However, there is little doubt that Molecular Genetics will in future complement the Quantitative Genetic approach to Animal Breeding in a range of important ways.

Chapter 3

Designing Performance Recording Operations

Keith Hammond

Introduction

Chapter 1 highlighted the importance of the information system to achieving genetic change in productivity and product quality, and the significance to Animal Breeding of the recent and anticipated developments in Information Technology. In Chapter 2 we briefly discussed the nine key decision areas and the three primary components of the Modern Breeding Approach. The approach depends on having cost-effective and reliable direct or indirect measurements for the traits in the breeding objective. Hence, the approach assumes that an effective performance recording system is operating, as the performance recording operations generate much, but not all, of the information necessary for the range of breeding decisions. This wider perspective, depicting both the primary information sources and the biological diversity which we aim to manipulate, along with the three operational components in the Modern Breeding Approach, is diagrammed in Figure 3.1.

What is Performance Recording?

In 1984 Brinks defined performance testing or record of performance as "the systematic measuring and recording of performance or indicators of performance traits. These records become a data bank and, upon proper manipulation and analysis are used in selection [really breeding] and management programs". This definition is fine but much operational detail is implied by the word 'systematic'.

A fundamental but often unstated reason for performance recording is to improve decision-making. Dommerholt (1987) has termed the information services used by producers to aid breeding and management decisions, Decision Support Systems. Increasingly, performance recording will be involved with **maximising the information content per unit of investment.** This means addressing:

- **What decisions are to be made and when?** By back-tracking from the necessary actions and the decisions associated with these, we are able to determine what information must be generated in the performance recording operations.

- **What results and output summaries are needed to aid these decisions?** By carefully considering the decisions and at what points and when in the breeding operations each is required, the most appropriate output in the most useful form and format will become clear.

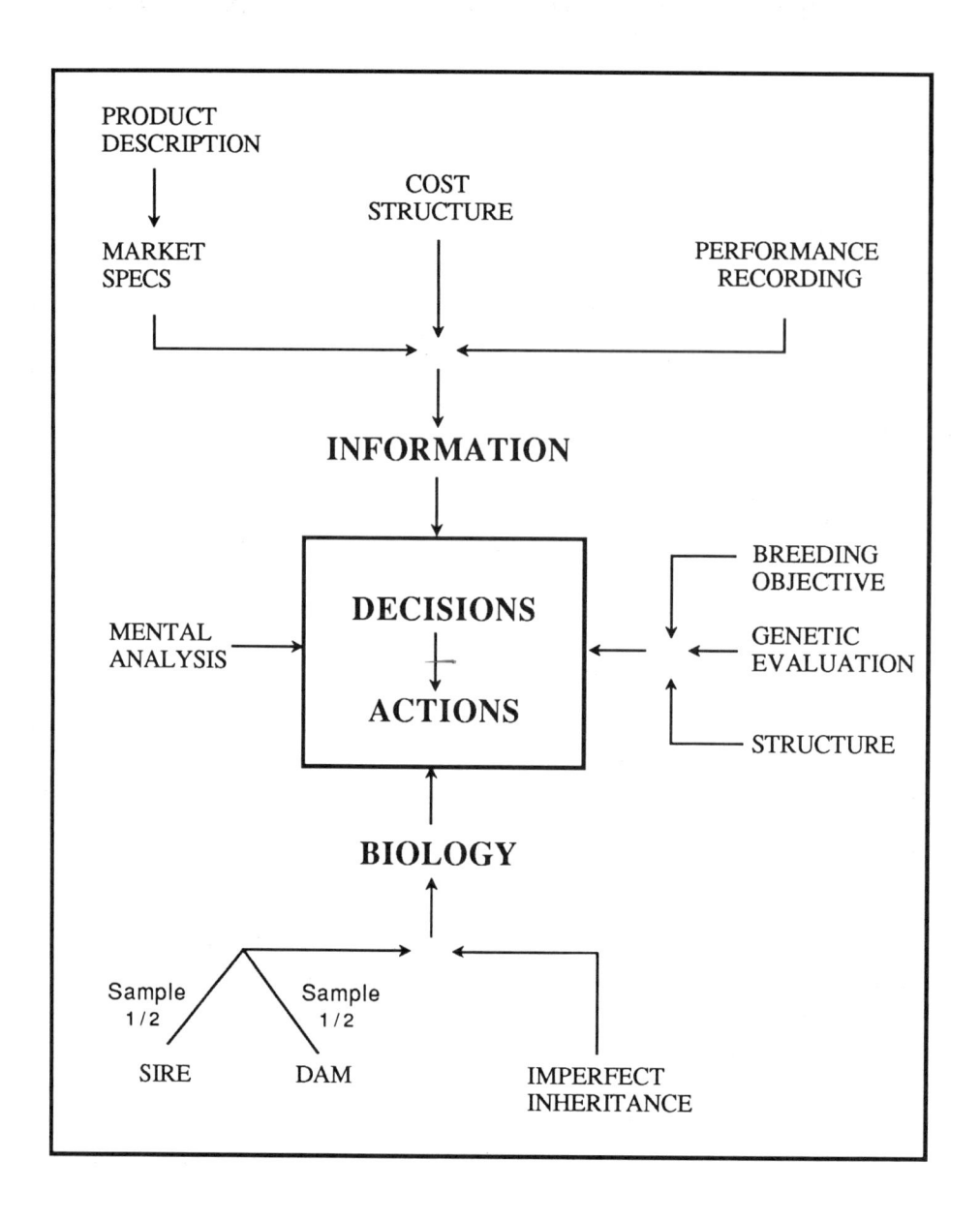

Fig. 3.1 Performance Recording and Animal Breeding.

- **What inputs are needed**? For example, which animals must be identified and what measurements recorded on each? Which identification system should be used? Which measurements, including visual scores are required, how to make them, e.g. weigh or score birthweight, and how frequently? What other data should be recorded, e.g. management groups, mating data?
- **How should all data be managed**? Data management covers a range of operations from measurement, through checking, transferring, storing, retrieving, and again editing the data; what analyses are required and when; and arranging for the regular analyses to present the results of these timely and in the preferred forms for use in the regular management and breeding decision-making processes.
- **How should the animals involved be managed to maximise cost-effectiveness, that is information content per unit of investment?** The most important item here is: How to structure management groups? Maximising the information content of performance recorded data is centred on the basics of good **experimental design.**

Performance recording:

- Is **the** primary supporting operation to Animal Breeding.
- Is commonly disregarded as being trivial - how wrong they are!
- Involves serious experimental design issues which are very important to the total genetic improvement operation, particularly those relating to management group structures, including cross-linkage between management groups to maximise information content.

Notice also that the definition of Performance Recording involves 'use in breeding **and management'.** Classically, Performance Recording has been associated with the breeding operation but increasingly the records themselves will become part of the total information system for the herd or flock, involving animal performance, accounts and other farm information to be used for the range of decision-making and other purposes. Figure 3.2 summarises this introduction to Performance Recording.

Increasingly, the performance records for a herd or flock will incorporate both on- and off-farm information - see also Chapter 6 on Within- *versus* Across-Herd/Flock Evaluation. Automation, computing, communications, and identifying, measuring and recording of animals and their product will become increasingly important to performance recording.

Now, to address each of the above listed five items.

What Decisions are Required, and When?

In planning performance recording operations it is common, but incorrect, to commence by considering animal identification and what to measure. Such a planning approach will generally be inefficient, ineffective, and tend to 'grow like topsy'. As mentioned previously, performance recording is being done to aid decisions in the breeding, other management and marketing areas.

WHAT IS PERFORMANCE RECORDING ?

The <u>systematic</u> measuring and recording of performance or indicators of performance, on- and off-farm

↓

| DATA BANK | ←------- Other production, economic and marketing data |

Manipulation | Analysis

↓

DECISION AIDS

for

- **BREEDING**
- **MANAGEMENT**
- **MARKETING**

The 'HERD INFORMATION SYSTEM'

or

'DECISION SUPPORT SYSTEM'

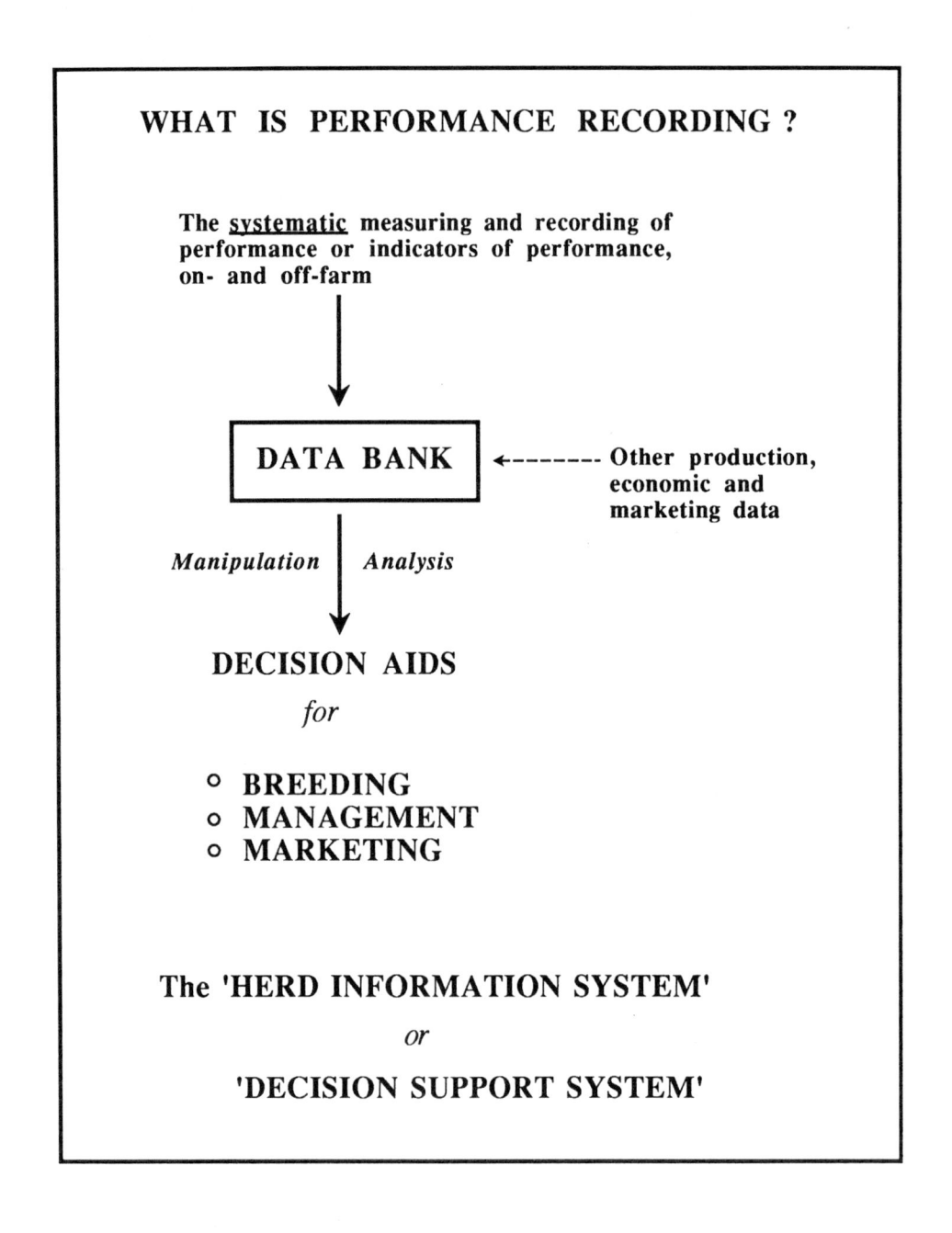

Fig. 3.2 Performance recording contributes to the Herd or Flock Information System for use in breeding, and in other management and marketing decisions.

Hence, **the logical starting point** in planning performance recording operations is to identify all decisions that are required throughout a year of operation, when they are made and where: in the office, shed, yard, field, sale-ring, etc. **A large Year Planner sheet** is a useful aid here for species such as sheep and cattle where the major operations such as mating, parturition and weaning only occur once or twice annually. For species with short reproductive cycles using continuous mating, the decision cycles have much shorter time frames.

These time frames are critical in the overall design of performance recording operations and particularly to the computing strategy used. For example, if decisions are made weekly to select potential replacement breeding stock and immediately prior to the unselected going to slaughter, as is done in some pig breeding operations, then analyses to produce EBVs need to occur within hours of the measurements being taken. With current technology this means either the breeder or a technician being in a position to analyse the data on-farm. Whereas in a dairy industry practising strictly seasonal calving the Breeding Value Summary for bulls being progeny tested is only required annually. Because the data for these dairy cattle evaluation analyses will come from a large number of herds, the computing is best conducted by a central bureau.

What Output is Required, and When?

For each of the decisions on the planner, the next operation is to determine precisely what **summary information** (results of analyses of the recorded data) is desirable, **when it is required** (timing must allow for collation, analysis and, where relevant, for turnround with the processing centre), the **form of output** (soft copy on computer screen or via a disk, hard copy on specially formatted paper or in a sale catalogue) and **the format** itself which maximises ease of use and minimises errors during interpretation.

Current Performance Recording systems of most industries still include too many results in each summary. Interim results are valuable for overall checking of analyses but when many of these interim results are regularly included along with final results the necessary decision processes can be hampered. This may even lead to reduced genetic change, e.g. through lack of understanding, the breeder uses a mix of the final and interim results. Of course, where hard copy is involved, it also adds to 'the paper war'. For example, the printing of EBVs for each trait for each animal and for its sire and dam is not necessary and can be replaced by a properly weighted single index value. This should be much easier to use and more effective than the breeder or buyer mentally computing the overall value - almost an impossible task to do correctly once more than two or three EBVs exist for each animal. A sound performance recording system should be capable of producing the more detailed summaries on demand. The move to simplify output will develop as the analytical procedures are improved and the confidence of seedstock breeders and buyers in these procedures increases.

The timing of output is also critical to maximising genetic gains. Output must be to hand when all potential animals are available for a particular decision process. If this does not occur both the intensity of selection and the generation interval, two important determinants of genetic change, can be seriously affected, or costs can be increased because of the extra time culls are held. Timing of the majority of breeding decisions will be determined by a small number of events primarily concerned with mating, e.g. age at first mating, frequency of mating, and date of mating; and also by events such as marking, weaning, time for turnround of reports from the central processing unit, and when seedstock are marketed.

The form and format of output is now changing rapidly with the many developments in computing and communications technology. In fact, these developments are enabling the collection of data from many more points, much better checking and editing of data, far more sophisticated analyses, e.g. introduction of BLUP procedures, more and better summarisation of results, e.g. single indexes rather than single trait EBVs, which are delivered more timely. In short, the many improvements underway to performance recording output form and formats are dramatically increasing the efficiency and effectiveness of the operations, with the result that genetic change in livestock populations throughout the world has generally increased during the 80s, in some cases dramatically so.

What Inputs are Required?

Once the necessary summary information and its timing is known, the next step is to determine what data needs to be collected and when. Again, this part of the planning is done in reverse mode, by first deciding on the necessary measurements, followed by which animals are to be measured, other data which must be collected, the animal identification system to be used, before reviewing the additional labour requirements. Several iterations may be required and of course the benefits and costs of various procedures will need examining. This will include evaluating the benefits to genetic improvement and from other managerial gains of each additional measurement, before finalising these inputs to the recording system.

What measurements?

A measurement may be taken to provide information to assist one or more of the management, breeding and marketing decisions already decided as important to the profitable operation of the herd or flock. The information required to aid these decisions should already have been identified on the Year Planner. Some measurements will need to be made on the same animals on more than one occasion, for genetic or non-genetic reasons.

Take special care to understand the complete relevance of measurements such as female reproductive performance, where higher reproductive rates also allow increased selection intensities, in addition to contributing to cash flow on the current herd or flock and to the breeding objective. Further, the higher the reproductive rate the more relatives and contemporaries, thereby also adding slightly to the accuracy of each EBV.

What makes a measurement useful?

Figure 3.3 summarises the factors that contribute to the value of a particular measurement. Note that this may well differ between seedstock producer and seedstock buyer, because they are likely to utilise the information somewhat differently. Hence, a producer of commercial product who buys all seedstock is not particularly interested in the heritability of a measurement whereas seedstock producers are. Buyers of replacement stock are particularly interested in measurements which are closely associated with profit from the **current** herd or flock and, for repeated measures such as weight of lamb weaned, with the repeatability of the measure. Buyers are further interested in making the measurement easily and at low direct cost only on the sex of interest, whereas breeders will generate more genetic information by being able to measure both sexes.

It is important to understand the **different uses of measurements**, and the implications of this for the statistical accuracy, freedom from bias, and precision, minimum random error about the measurement. When measurements are used for the purposes of transaction of animals or product, normally only one measurement of particular traits, e.g. fat colour and body weight, will be made and payment for the animal or product will be determined by these measurements of the traits and the price schedule.

On the other hand, when measurements are used to aid breeding decisions the primary consideration is whether the information generated is identifying genetic differences between the animals measured, i.e. what is the extent of this variation and the heritability of the measure? Further, and particularly when the modern genetic evaluation procedures are being used, it is rare for a single measurement to be used on its own in genetic comparisons. The BLUP analysis carried out on the data will normally combine many measurements across repeated records on the same animal, different traits on the same animal and across all the relatives and contemporaries of all animals, to produce the final EBVs for each animal. Hence, the value of a measurement must be judged according to its use. This concept is not at all well understood.

How to take measurements

Measurements may be made mentally, where no equipment is used, i.e. visual observation; or manually, where some piece of equipment is used to assist with the measurement, e.g. rule, weigh scale, real time ultrasonic scanning machine; or automatically, where automated identification, animal control, and measuring devices are combined to overcome the need for human intervention.

Again, care is needed in interpreting the value of mental *versus* manual *versus* automated measurements. For breeding purposes a mental measurement may be all that is possible with current technology, e.g. a score for ease of calving or for body conformation. For those traits which are reasonably highly variable, a mental score may actually detect 80% or 90% of the genetic differences between animals that are detectable using equipment and additional labour inputs, **providing** the scoring system is carefully designed and used. For breeding purposes, the interest is in whether a particular measurement will make a significant contribution to the accuracy of the selection index of all measures used, i.e. to the association between the index of measures used and the breeding objective.

WHAT MAKES A MEASUREMENT USEFUL ?

A. TO A SEEDSTOCK BREEDER ?

    1. Variable ↑

*   2. Heritable ↑

*   3. Correlated with economic traits ↑

*   4. Taken prior to reproductive age

*   5. Taken on both sexes

(Uptake is simpler if it can be taken easily and cheaply)

B. TO A SEEDSTOCK BUYER ?

    1. Variable ↑

*   2. Correlated with current profit ↑
       (irrespective of heritability)

*   3. Repeatable
       (for repeated measures) ↑

*   4. Taken early in life ↑

*   5. Taken on the sex of interest

*   6. Low cost + Easy ↑

* = Differences   The more (↑) the better

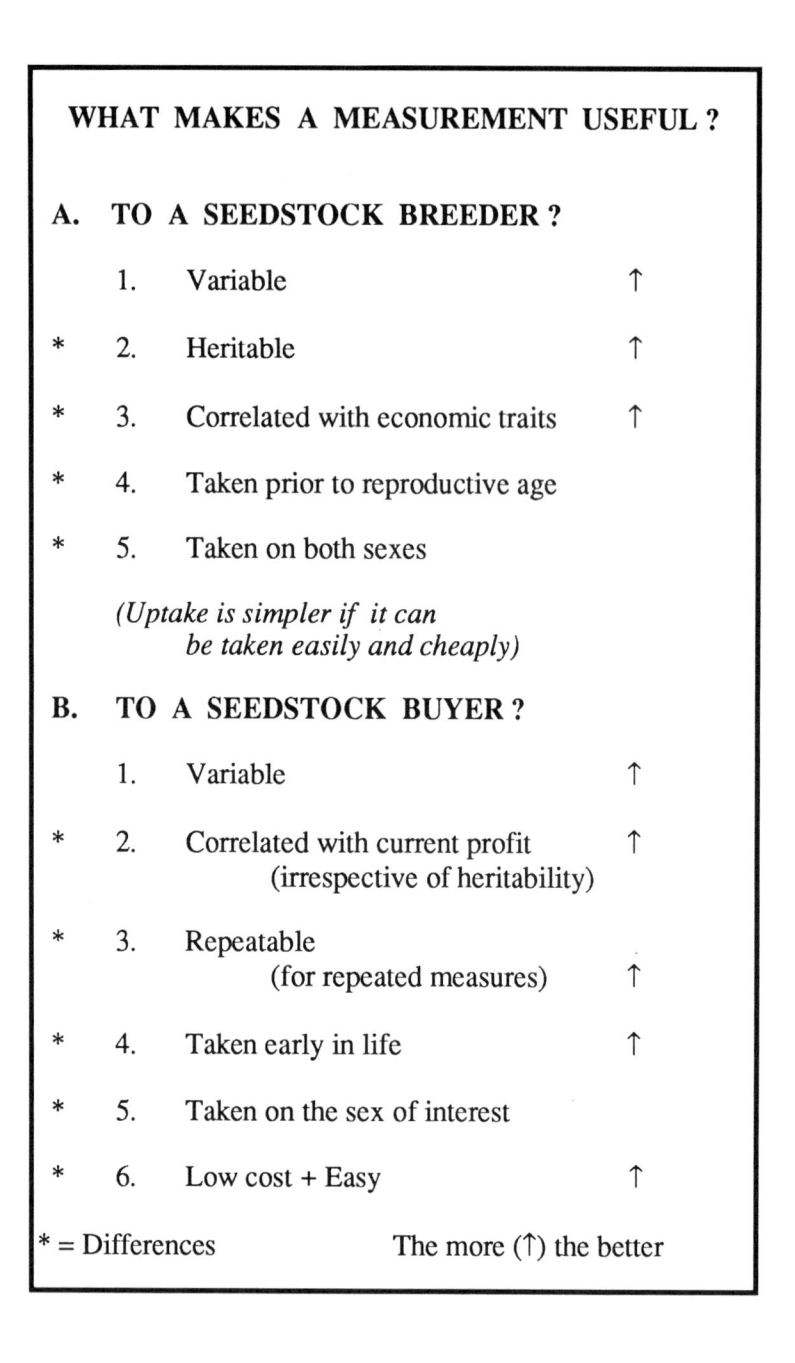

Fig. 3.3 The value of a measurement to breeders and buyers of seedstock

Figure 3.4 provides an approximate, generalised rating for mental, manual and automated systems of measurement, based on a number of criteria associated with the value of the information generated and the nature of the system, i.e. its reliability, flexibility, current usage and cost. This information is meant as a guide only, to encourage correct thinking on the value of measurement systems.

Irrespective of the type of measurement system used, a **standard protocol** should be designed for taking each measurement. This may include the type of equipment, its preparation at the outset and cleaning and standardisation prior to the next round of measuring. The unit and precision of the measurement should be clear. With subjective measurements the number of categories and the definition of each must be defined.

Special care is required to standardise subjective, categorical measurements. Guidelines now exist for most industries and the more common subjective measurements. If possible, the minimum number of categories used should be four or five as this will help during analysis to detect the genetic differences between animals. All categories should be defined to be mutually exclusive, i.e. a measurement for an animal must fall into one or other category. At times this can be difficult for traits such as ease of calving when malpresentations are also included, but it is not impossible. Finally, when making these subjective measurements, it is most important that the complete scale be used - placing all animals in the one category means that no genetic differences can be detected! The human mind tends to play tricks when using categorical scoring and use fewer categories than it should.

Direct *versus* indirect measures

When the trait measured is included in the breeding objective, i.e. it is directly involved in generating costs or returns, it is known as a direct trait and measure. When it is not possible to measure one or more direct traits in the objective, other measurements may be taken to help increase gains in profitability. These other measurements are known as **indirect** or **secondary** or **marker** traits. Note, when the marker trait is measured by molecular means these are the so-called Molecular Markers that are currently creating so much interest. These then are simply another form of indirect measurement of a trait in the objective.

Indirect measures are used where the direct traits are:

- Costly or difficult to record on sufficient animals, e.g. food intake, reproduction, disease resistance, meat yield and quality.
- Not expressed early in life, e.g. female reproduction, milk production and survival in the herd or flock.
- Sex limited, particularly where reproductive rate is low, e.g. female reproduction and milk production in cattle.
- Not highly heritable but variable and correlated with the marker trait.

COMPARING MEASUREMENT SYSTEMS			
	Mental ↔	**Manual** ↔	**Automated** ↔
INFORMATION:			
• Amount	+	++	++++
• Accuracy	+	++	+++
• Analysis	–	+	+
(EBVs + Trends)			
• Timing	++++	+	++
SYSTEM:			
• Reliability	– → ++	+ → +++	→ +++
• Flexibility	+	+++	++++
• Occurence	+++	++	–
• Cost	++++	+	++
	The more +s the better / more common		

Fig. 3.4 General ratings for the broad categories of measurement systems

There are many examples of indirect measurements, e.g.
- Growth rate and body weight as measures of feed efficiency in the growing animal and mature female.
- Scrotal circumference in cattle as a measure of calving rate of the mates, calving rate of the daughters and calving rate of the mates of sons.
- Birth weight and pelvic diameter as measures of ease of calving.
- Fat depth as a measure of total fat in the carcase.
- Muscling score as a measure of meat yield.
- Dairy character as a measure of milk yield.
- Conformation as a measure of overall fitness in the production system.

In addition, there are logistical difficulties for the seedstock producer trying to obtain sufficient genetic information on such traits as meat quality, for seedstock producing herds and flocks slaughter few animals. The difficulty of following animal identification from original herd or flock to the point of measuring, highlights the need for indirect measures of carcase quality on the live animal.

Measuring locations

With developments in computing and communications, measurements taken over time and at several different locations are increasingly being combined, e.g. combining measurements from the same animal and its product taken within the herd or flock, at central test stations, at feedlots, at abattoirs and at central measurement laboratories for such products as milk and fibre quality. Figure 3.5 shows this and highlights the central importance of within-herd/flock input for the total breeding operation.

Which animals to measure

With the modern genetic evaluation procedures the more animals which can be measured, the better, as all information is used in the simultaneous analysis and all animals receive EBVs and contribute information to the calculation of genetic and non-genetic trends. Molecular and engineering technologies will increasingly enable direct or indirect measurements of each trait in the objective, on each animal of each sex early in life.

Having said this, the collection of information on most measurements involves additional costs - can you think of examples where no additional cost is involved? Hence, it is important to evaluate the contribution made to each decision by the measurements taken on the different groups of animals in the herd or flock structure.

Sometimes a small group of animals will be measured with equipment, e.g. a weight at birth and used to standardise a subjective measurement, e.g. categorising calves at birth into five weight groups, viz. tiny, below average, average, above average and large.

What other data should be collected?

It is important to determine exactly what other data needs to be associated with each measurement e.g. animal and parent identification, date of measuring, management group, location, scorer, and data required for the analysis e.g. mating data, body condition. Again, the protocols for these data must be clearly defined, particularly the units to use.

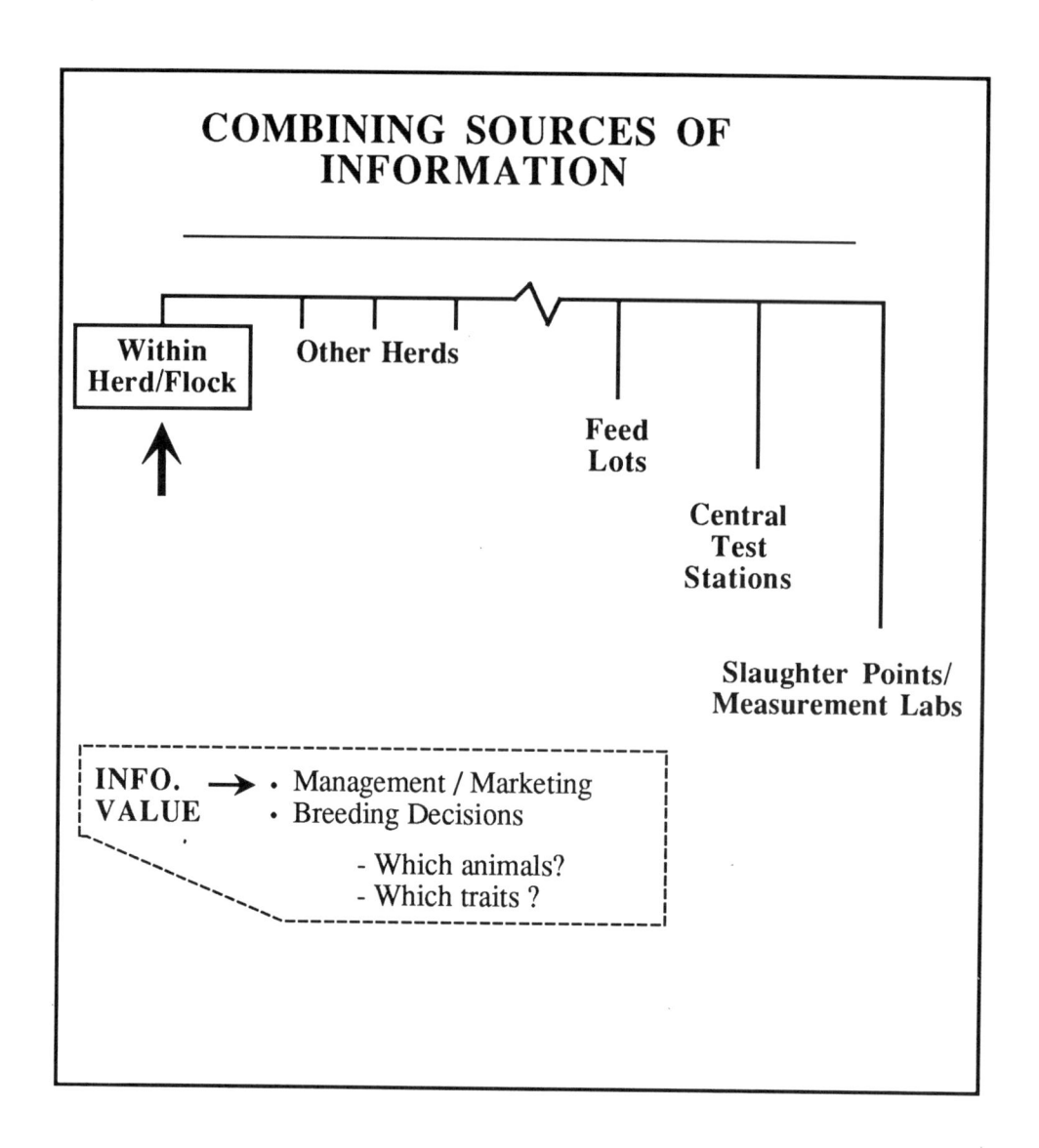

Fig. 3.5 Increasingly, data relating to the same animal collected over time and space are being combined

Which identification system to use

Animal identification is required wherever records on animals are taken over time and space before breeding decisions are taken. It is necessary to determine when and whether all animals, males and females are to be identified, and the identification system to be used.

No single manual or automated identification system is cost-effective for all purposes, and backup systems may be required to overcome losses of tags or inability to read various individual identification marks on animals.

If different identification procedures are used at different stages of the lifecycle, or between live animal and carcase, milk sample or fleece sample, the codes for each animal must be tied together (merged) preferably when first used.

Where sires are used across herds or flocks the identification system must enable each sire to be identified as the same individual at each location. Failure to do this greatly complicates future use of the data in across-herd/flock genetic evaluation analyses.

Decisions must be made on the level of parentage mis-identification which is acceptable, particularly where mis-mothering is occuring. However, it is unlikely that such mis-mothering will be so severe as to prevent genetic progress for most traits of interest. Providing the mis-mothering events are reasonably random for the traits of interest then they will simply add to the 'noise'. This 'noise' will lower the heritabilities but generally this reduction will not be great.

What added labour requirements are needed

Recording the performance of animals and breeding operations place additional demands on labour, often both quantity and quality. Commonly, there exist many opportunities to reduce these additional requirements, providing the options are carefully evaluated, in relation to their impact on the efficiency and effectiveness of the recording and breeding operations. This is considered further in Chapter 20 covering Design of Straight-Breeding Programs where Figure 20.1 lists the operations contributing to the differences between herd or flock in the type, amount and quality of labour used in recording and breeding.

What management system

Advances in computing and communications technologies have resulted in central data management system design for herds, flocks or breeds becoming a highly specialised operation. Bureaus and software houses specialising in data management systems for performance and other records for both central and on-farm data management now exist for many of the livestock industries. Attitudes tend to vary between industries on the willingness of individual breeders to utilise central processing bureaus, e.g. compare the dairy cattle and beef cattle industries of most countries with the pig industry of at least some countries where individual breeding enterprises prefer to retain total control of their whole data system. This has important implications for system design strategies and for whether the industry will support across-herd/flock genetic evaluation.

However, there will be many manual aspects of data management on the farm and at other locations where data are to be collected that will also need careful definition and design to ensure that the data collected throughout the breeding operation are accurate, timely and collection is done at least cost.

Care is required to incorporate checks throughout the data management system, for identification, measurements being within range, other data required are also collected and within range, the whole system is adequately backed-up to prevent major losses, and appropriate security protocols are incorporated against unauthorised human interference and 'viruses'.

Managing Stock to Maximise Effectiveness of Recording

The most important aspects here are the size and structure of management groups. These are utilised in the genetic evaluation analyses to make direct and fair comparisons. If animals are treated differently and this is not recorded then to the extent that these different treatments influence the expression of each trait being measured, the resulting EBVs will be biased. All differential treatment of animals or groups of animals should be recorded even if some of this information is not currently used in the EBV analyses.

In addition, genetic comparisons between management groups and between herds or flocks for across-herd/flock genetic evaluation analyses are made by reference to parental links, e.g. the same sires being represented in each management group. To maximise the information content of the records it is useful then to have sires represented across as many management groups as possible. In addition, sires should be rotated around mating groups between matings or, with continuous mating such as is used in pigs, sows should be mated to different boars at each successive mating.

When splitting groups of progeny ensure that two or more sires are represented within each group and split on sexes and ages of progeny and then if necessary split on age of dam.

Maintain management groups as large as practicable. In beef cattle for instance ultimate contemporary group size for an analysis can become very small, sometimes with only two or three animals being compared directly in the one group and all from the one sire. This will add negligable information to the genetic evaluation for the herd or flock. Aim for management groups of 20 to 30 or more.

Planning the whole Performance Recording Program

It should be obvious by now that systematic planning of Performance Recording operations is required to produce maximum genetic information for the herd or flock at least cost. Given the importance of the recording operation to the breeding program and the many opportunities for inefficiencies to be introduced at the outset and subsequently, planning should be taken seriously, and a plan **documented** and updated from time to time. This rarely occurs!

Evolution of the program

Performance Recording operations must evolve. The correct evolutionary path must:

- Be technically sound, otherwise substantial time and money will eventually be wasted in overcoming the technical deficiencies.
- Promote industry uptake, feedback and learning. Here the current and anticipated future industry structures are important considerations.
- Involve a minimum of backtracking - once users of a service are given some concept or piece of information it is difficult for it to be withdrawn.
- Provide for system upgrades to slot in easily both in the field and at data processing centres.
- Contribute as much data as possible to the research and development program for the service.

Further Items

Recording is not a spare time job

It is vital that time be allocated in the daily and weekly work schedule for recording purposes, particularly for the office procedures as these are often the first to suffer.

GIGO operates

Remember, Garbage-In Garbage-Out! Accurate recording of animal identification, management groups, and accurate measuring and recording of each trait together with other associated data is necessary for obtaining the best EBVs. Remember, the BLUP procedure simultaneously utilises the majority of the information contained in the complete data set, i.e. all years of records, all animals and measurements are analysed simultaneously. Errors in pedigree and in other parts of animals records, unintentional or otherwise, will impact throughout the analysis, year after year, the extent of this depending upon the amount of use made of the animals whose records are incorrect. Of course as the records become ancient they will have less influence.

Artificial Breeding and Performance Recording

Extra care in recording is needed where combinations of AI and natural mating are used. When embryo transfer is used both the donors and recipients should be recorded. Note the relative value to genetic evaluation of the different possible types of recipients. Nondescript recipients reduce the information available to evaluate donors, their mates and the offspring. With split embryos it is necessary to identify each split to an embryo. Gene segregation has occurred between embryos but two splits of the same embryo are identical. Figure 3.6 summarises the data required to monitor and correctly evaluate offspring resulting from combinations of natural and artificial reproduction. It should now be obvious that male and female artificial breeding together may substantially complicate a recording system!

Graduating from Within- to Across-herd/flock Recording

Where a major portion of the matings in an industry utilise AI across-herd/flock genetic evaluation is now the norm e.g. most diary industries. Where this is not so then special operations such as Reference Sire Schemes must be introduced to obtain valid across-herd or flock genetic evaluation - see Chapter 6. In these situations it is preferable that breeders learn the requirements of modern recording and genetic evaluation systems within their own herd/flock before their data is used in across-herd/flock evaluations. This is possible in some industries but not in others.

The primary issues for across-herd/flock recording and evaluation with low AI usage are:

- Developing and maintaining a central organisation for the across-herd/flock recording to ensure that regular and useful genetic evaluations are produced. Droughts and other difficult times cause particular problems.
- Developing strong genetic links via some common sire usage - how many progeny per reference sire in each herd or flock? Ensure linkage across herds or flocks is strong enough for each trait that is to be evaluated. Winning universal acceptance of proposed Reference Sires from all involved is not always easy, and this acceptance impacts on the degree of linkage.
- Ensuring all females involved, Home and Reference Sires are mated to representative groups of females - in the first few rounds of mating this remains important even when animal model BLUP procedures are to be used for the genetic evaluation analyses.
- Ensuring the identification system enables the same sire to be identified in each herd or flock that is used.
- The perceived importance of genotype by environment and sire by herds or flocks interactions for each of the traits being evaluated.
- Recording and evaluating of animals introduced from external herds/flocks and other countries and of grade and crossbred animals.

How many EBVs?

Often there is discussion in the industry over the number of EBVs which can or should be computed by service bureaus and analytical software. However, a recording and evaluation service or package should possess the capability to evaluate measures of all economic traits for each major sector, production system and market of an industry; and possess the flexibility to enable particular users of the service or package to obtain only the EBVs they require.

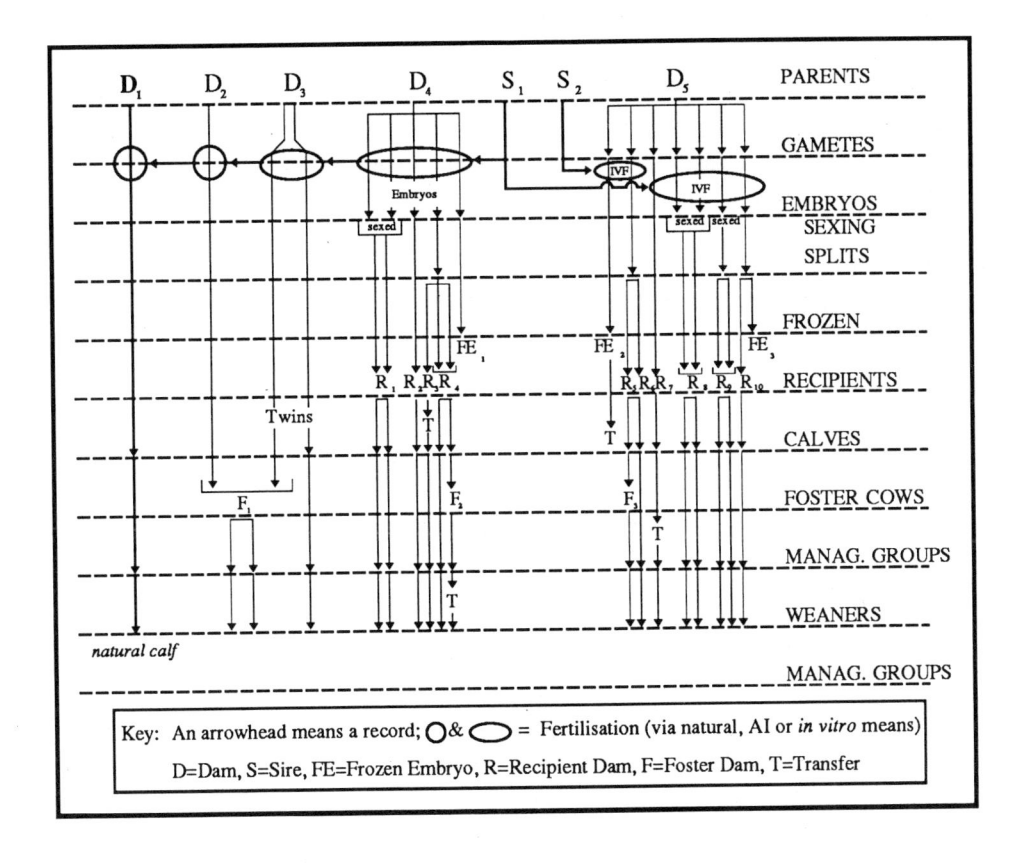

Figure 3.6 The recordings required to monitor with-in herd or breed society and correctly evaluate offspring resulting from combinations of natural reproduction and embryo transfer and splits, In Vitro Fertilisation (IVF), Sexing and the use of Foster dams

Dos and Don'ts in Performance Recording

The don'ts
- Dont't measure a few, central test a couple.
- Don't otherwise progress with no formal plan.
- Don't identify or record only replacements.
- Don't change identification during the life of each animal.

- Don't replace all sires each year/mating.
- Don't use one male over all heifers/hoggets/gilts only.
- Don't cull some prior to recording the first measurement.
- Don't have single sire Management Groups.
- Don't have small Management Groups.
- Don't include unlike treated animals in the one Management Group, and otherwise unfairly treat parents or offspring.
- Don't use partially dislectic people in collection or transfer of data
- Don't treat recording as a "spare-time job".

The dos - approach
- Custom design each herd or flock for cost-effectiveness.
- **First** determine what decisions are to be aided, and when, and use a PLANNER sheet:

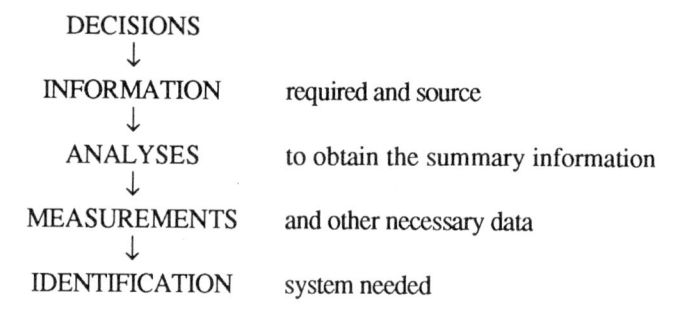

$$\begin{array}{ll} \text{DECISIONS} & \\ \downarrow & \\ \text{INFORMATION} & \text{required and source} \\ \downarrow & \\ \text{ANALYSES} & \text{to obtain the summary information} \\ \downarrow & \\ \text{MEASUREMENTS} & \text{and other necessary data} \\ \downarrow & \\ \text{IDENTIFICATION} & \text{system needed} \end{array}$$

- Consider: Recording all females, particularly where reproduction is of interest, and include mis-matings. A Breeding Female Inventory recording system is often superior to an Offspring Inventory.
- Consider: Recording status (fate, etc.) for all females.
- Build checks throughout the data system, but accept an overall error rate of 2-5%.
- Provide also for future further development of the recording operations.
- Stage implementation, if necessary.
- Prepare a written plan; by year/lifecycle, covering all animals: males, castrates, females.

The Dos - Detail
- Record all sires used, natural + AI, and when, for each mating.
- Time AI to coincide with natural mating, or include a sire from natural mating in AI program.
- Mate Reference Sires to produce at least 20 progeny.
- Plan compact calvings/lambings.
- Rotate sires between matings.
- Record all donors, recipients, fosters, and use previously recorded females as recipients.

- Identify all embryos, and splits back to embryo.
- Mother-up offspring to dam, if possible.
- Record all animals/groups treated differently (= Management Groups).
- Otherwise maximise size of Management Groups.
- Maximise sire spread across Management Groups.
- Where Management Groups are further divided, split first on sex and age, then on dam age only if necessary.
- Carefully plan timing of all recording and processing operations to ensure the necessary information is available prior to decision-time.
- Gain recording experience within the herd or flock before being involved in across-herd/flock genetic evaluation.
- Utilise only committed people in the recording process.

References

Brinks JS (1984) The concept of performance testing. In: Breeding Beef Cattle in a Range Environment. Fort Keogh Research Symposium (Ed. T.C. Nelsen and R.A. Bellows). (USDA-ARS Miles City, Montana). pp. 109-113

Dommerholt J (1987) National cattle data base and on the farm automation. In: Automation in Dairying. Proceedings of the Third Symposium Organized by IMAG Wageningen, The Netherlands. pp. 345-356

PART II : Genetic Evaluation

Chapter 4

Principles of Estimated Breeding Values

Brian Kinghorn

Heritability and EBV or EPD

What causes an exceptional animal to be so much better than its contemporaries? There are two basic reasons:

- It has probably inherited better genes making it genetically superior.
- It has probably experienced a better environment, through good luck.

In seeking genetic change we are not really interested in how much environmental luck an animal has had, because that source of superiority cannot be transmitted to the next generation. We want to be able to choose the animals with better genes.

The Breeding Value (BV) of an animal is a description of the value of an animal's genes to its progeny. We cannot see what genes an animal carries, so we can never fully **know** what an animal's BV is. However, we can estimate it. We can calculate an animal's Estimated Breeding Value (EBV) from various sources of information, including the following:

- Its own performance or phenotype for the trait of interest.
- Its own performance for other traits.
- The performance of its relatives for both the trait of interest and other traits.

How do we predict an animal's breeding value given its phenotype for a single trait most usefully expressed as a difference from the population mean? We must penalise for the good luck an animal is expected to have had. We do this by multiplying its phenotypic superiority (P) by the heritability (h^2) of the trait concerned, as seen in Figure 4.1.

$$EBV = h^2 P$$

In some countries, animal breeders use Estimated Progeny Differences (EPDs) rather that EBVs to describe the breeding merit of candidates for selection. Here is the difference:

- EBVs describe the value of an animal's genes to its progeny, and yet it only transmits half of its genes to any one progeny. So, given mates of average value, the predicted merit of an animal's progeny is half of its EBV. Alternatively, the predicted merit of the progeny of a mating pair is the average of their EBVs. This will be described in more detail later.

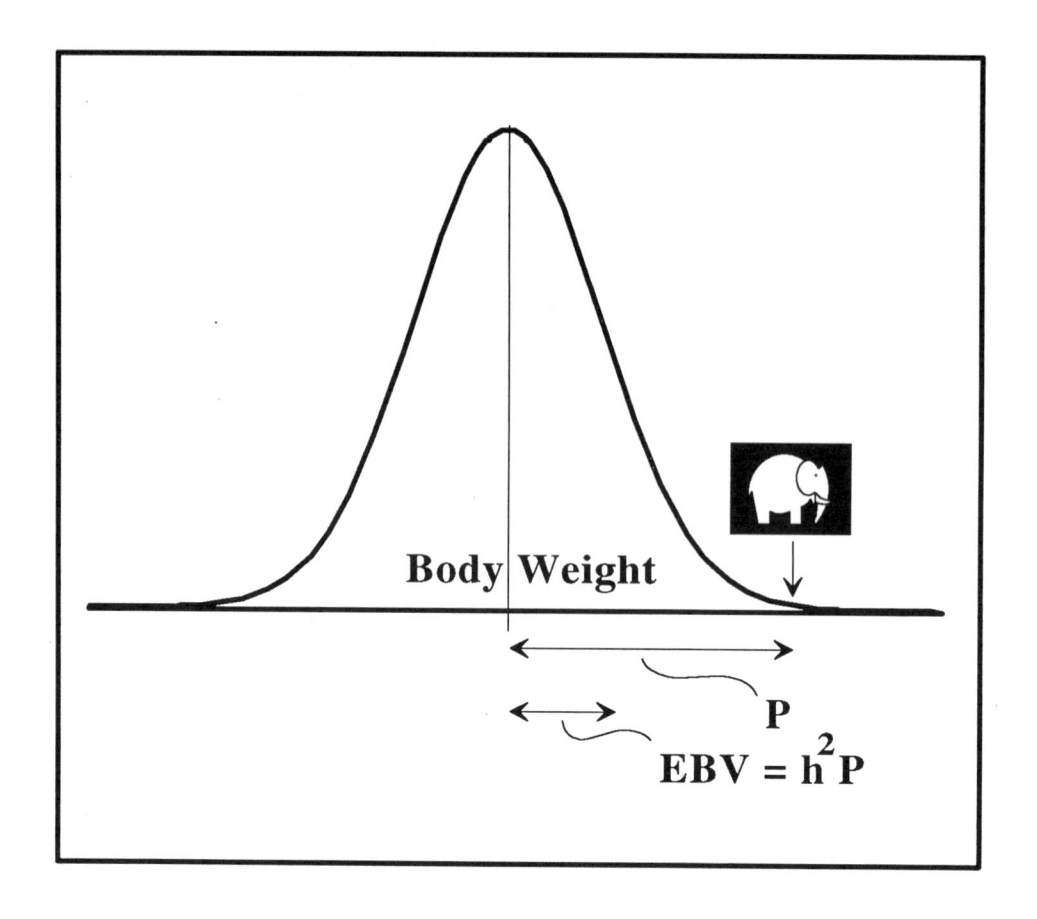

Fig. 4.1 A high performing animal has its phenotypic superiority over the herd or flock mean (P) penalised by heritability (h^2) in order to estimate its breeding value (EBV = h^2P) from its phenotype.

- EPDs have the half already built in. So, given mates of average value, the predicted merit of an animal's progeny is its full EPD. Alternatively, the predicted merit of the progeny of a mating pair is the sum of their EPDs.

Thus the relationship between EBVs and EPDs is quite simple:

$$EPD = \frac{EBV}{2}$$

In Figure 4.1, heritability is taken as 40%, or 0.4, which is in the high end of the range we find for commercially important traits, as can be seen from the heritabilities given in Table 4.1.

Table 4.1 Some heritability estimates for commercial traits[a]

Species and trait	Percent heritability	Species and trait	Percent heritability
Beef Cattle		Thoroughbred racing	
Calving interval	10	Log of earnings	50
Age at puberty	40	Time	15
Scrotal circumference	50	Pacer: best time	
Birth weight	40	Trotter	15
Weaning weight	30	Log earnings	40
Post weaning gain	45	Time	30
Yearling weight	40		
Yearling hip frame size	40	**Poultry**	
Mature weight	50	Age at sexual maturity	35
Carcase quality grade	40	Total egg production	25
Yield grade	30	Egg weight	40
Eye Cancer	30	Body weight	40
		Shank length	45
Dairy Cattle		Egg hatchability	10
Services per conception	5	Livability	10
Birth weight	50		
Milk production	25	**Sheep**	
Fat production	25	Number born	15
Protein	25	Birth weight	30
Solids-not-fat	25	Weaning weight	30
Type score	30	Post weaning gain	40
Teat placement	20	Mature weight	40
Mastitis susceptibility	10	Fleece weight	40
Milking speed	30	Face covering	55
Mature weight	35	Loin eye area	55
Excitability	25	Carcase fat thickness	50
		Weight of retail Cuts	50
Goats		Fibre Diameter	50
Milk production	30		
Mohair production	20	**Swine**	
		Litter size	10
Horses		Birth weight	5
Withers height	45	Litter Weaning weight	15
Pulling power	25	Post weaning gain	30
Riding Performance		Backfat probe	40
Jumping (earnings)	20	Carcass fat thickness	50
Dressage (earnings)	20	Loin eye area	45
Cutting ability	5	Percent lean cuts	45

a Adapted from Taylor & Bogart, 1988

Note that heritabilities can differ between populations:

- Some populations are in a more heterogeneous environment, reducing heritablility.
- The same trait in different environments may actually act as two traits - e.g. growth may depend on appetite in one environment and on efficiency in another.
- Different populations have different genes and/or gene frequencies, giving room for different amounts of genetic variation.

Here is an alternative view of this same penalising approach to estimating breeding values. Phenotype (P) expressed as differences from the herd or flock mean, are regressed by heritability, h^2. An animal with a +40 phenotype has an Estimated Breeding Value of +10, assuming a heritability of 0.25:

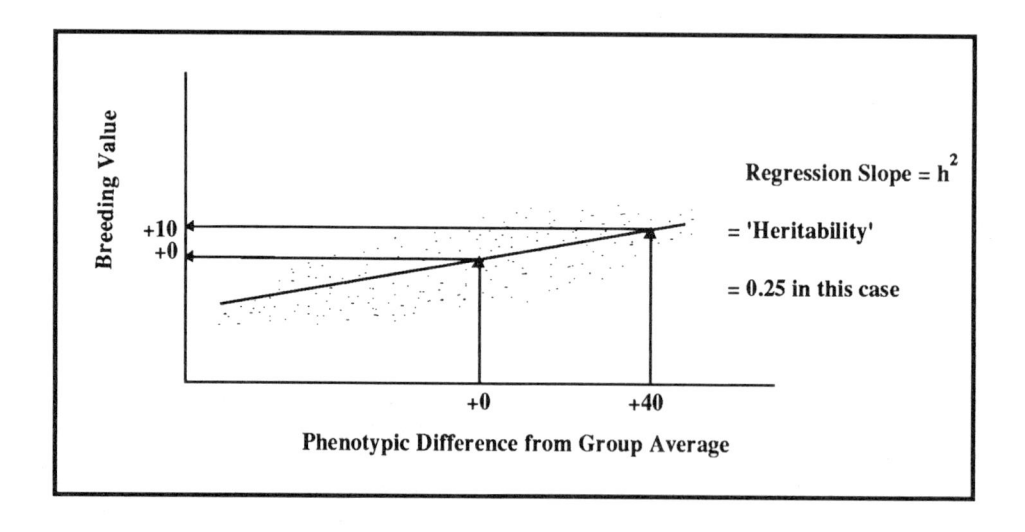

Fig. 4.2 As with figure 4.1, a high performing animal has its phenotypic superiority over the herd or flock mean (P = +40) penalised by heritability (h^2 = 0.25) in order to estimate its breeding value (EBV = h^2P = +10) from its phenotype.

Using Estimated Breeding Values

EBVs have a very useful property: The predicted merit of progeny is simply the average of the EBVs of the two parents used. If a ram has an EBV for wool weight of +1 (as a deviation from unselected rams), and a ewe has an EBV of +0.5, any progeny they have are predicted to have a superiority of 0.75 above progeny from randomly selected parents. Of course, this prediction rarely proves to be exactly correct, as described later, but if the EBVs are sufficiently accurate, the average value of many progeny will be about 0.75 superior.

Because EBVs can be averaged in this way, they are additive, and are thus often denoted by the letter $\hat{A}$. The hat (^) denotes this is a prediction or estimate of true breeding value A. Thus the predicted value of offspring from a mating is:

$$\hat{P}_o = \frac{\hat{A}_m + \hat{A}_f}{2}$$

where **o** is for offspring, **m** is for males and **f** is for females.

An example - Yearling Weight in Beef Cattle

Assume the average yearling weight is 300kg and $h^2 = 0.4$. Assume, for now, no effect of sex on yearling weight. Bull X is 340kg at one year. What is the average value of his progeny expected to be if he is mated to a representative group of cows?

$$P = 340 - 300 = +40\text{kg} \qquad \hat{A} = h^2 P = 0.4 \cdot (+40) = +16\text{kg}$$

$$\hat{P}_o = \frac{\hat{A}_m + \hat{A}_f}{2} = \frac{+16 + 0}{2} = +8\text{kg}$$

Note '0' for the representative cows. +8kg is a deviation from the average, giving a predicted progeny average of 308kg.

If he were mated to a top cow weighing 330kg ...

$$\hat{P}_o = \frac{\hat{A}_m + \hat{A}_f}{2} = \frac{+16 + (.4 \cdot +30)}{2} = +14\text{kg}$$

+14kg is a deviation from the average, giving a predicted progeny average of 314kg.

Note: not all offspring will be 314kg !!!:

- There is **genetic variation** within families. You are not identical to your full-sibs.
- There is **environmental variation** of 2 types
    - The whole progeny mean deviates systematically (droughts, good seasons, etc.)
    - The environmental deviations, elements of luck, for the progeny could tend to be high or low by random chance.

Dollar EBVs

As performance must be measured on a number of different scales, e.g. kg wool weight, micron fibre diameter, it is most convenient to express these all on the single scale of dollar value, or other economic units. Thus the dollar EBV (EBV$) of an animal relates to its overall effect on all important traits expressed in its progeny.

EBV$s or EPD$s then can be used to rank animals for selection. However, the power to rank with confidence depends on the amount of information available. If only tag numbers are available, then there is no useful information, and all animals are perceived to have the same average value. This is indicated by a zero width in the distribution of dollar EBVs shown in Figure 4.3.

With greasy fleece weight (GFW) measurements, there is some power to distinguish animals of high and low dollar EBV. However, where fibre diameter (FD) plays an important role in the breeding objective, GFW alone does a poor job of identifying the most valuable animals, and the dollar EBV distribution is quite narrow. As more information is recorded, there is increasing confidence in identifying animals of exceptional dollar EBV, these animals being represented at the right-hand-sides of the wider distributions in Figure 4.2.

Use of Dollar EBVs

Even when calculated from a selection index calculation or a BLUP (Best Linear Unbiased Prediction) analysis, EBVs still have that very useful property: The predicted merit of progeny is simply the average of the EBVs of the two parents used. This simple predictive property of EBVs can also be used to show the effects of selection intensity, which is higher when fewer, more elite parents can be selected, and level of measurement on response to selection for the whole flock.

In Figure 4.4, a smaller proportion of rams than ewes can be selected for breeding, contributing to their high mean EBV. The other factor, in this case, is the greater amount of information used to calculate ram EBVs, reflected by a wider distribution of EBVs for rams. Notice that the predicted merit of progeny is simply the average, or half-way-point, between the selected rams' and ewes' mean EBVs. The width of the EBV distribution of the progeny depends on how intensively they are measured.

Animals can be simply ranked on EBV$ values and the best selected for breeding. However, some traits are influenced not just by the genes of the animals expressing the trait, but also by the genes of their mothers. These effects are both expressed in the same measure - the performance of calves. An example is the influence of growth genes in calves and milk genes in their mothers, both contributing to the expression of growth in the calves. These groups of genes affect the direct sub-trait and the maternal sub-trait respectively. If EBVdirect (EBV_d) and EBVmaternal (EBV_m) are both available in a genetic evaluation report, then the strategies outlined in Table 4.2 can be used:

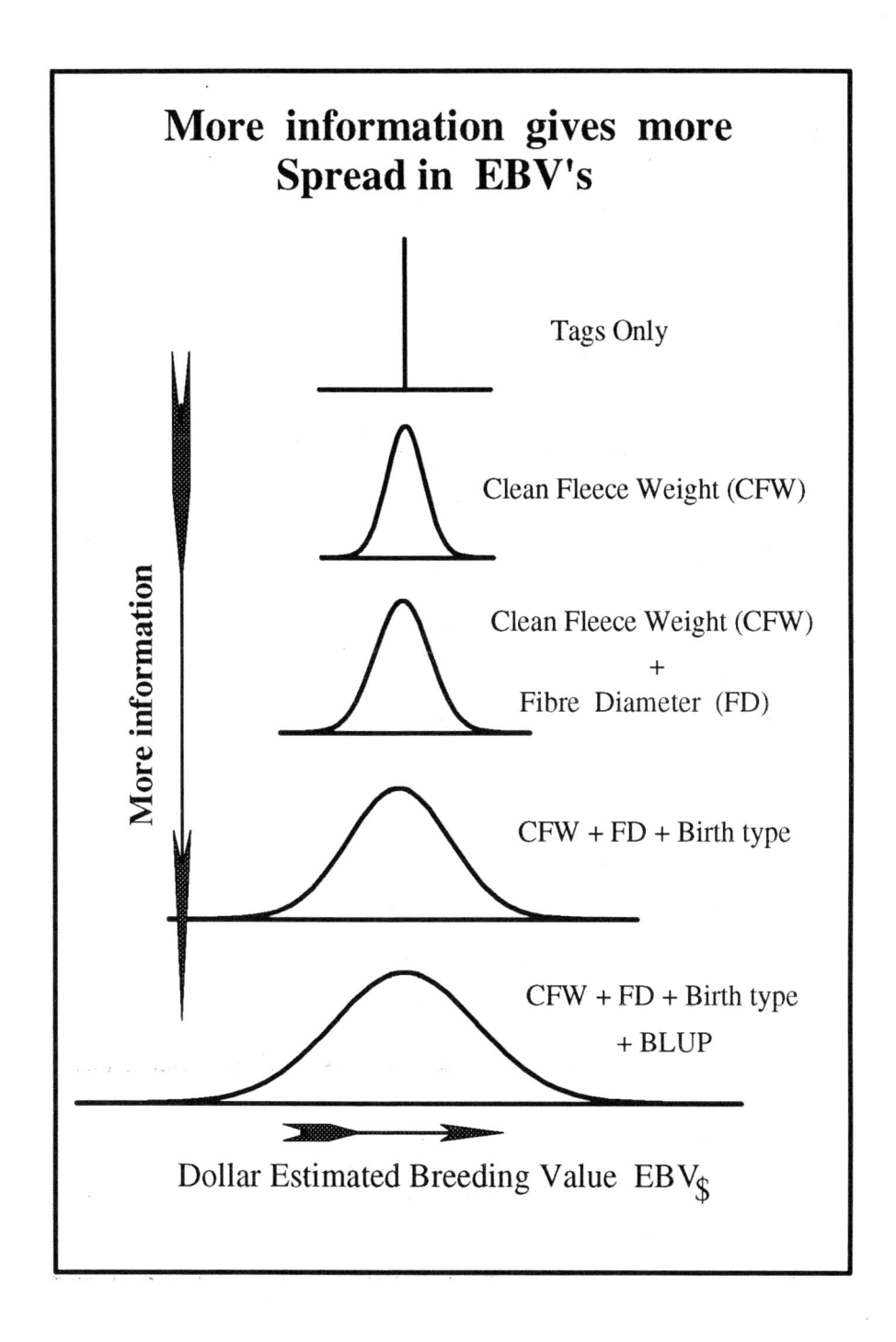

Fig. 4.3 The relation between the amount of information recorded on animals and the spread in their EBVs.

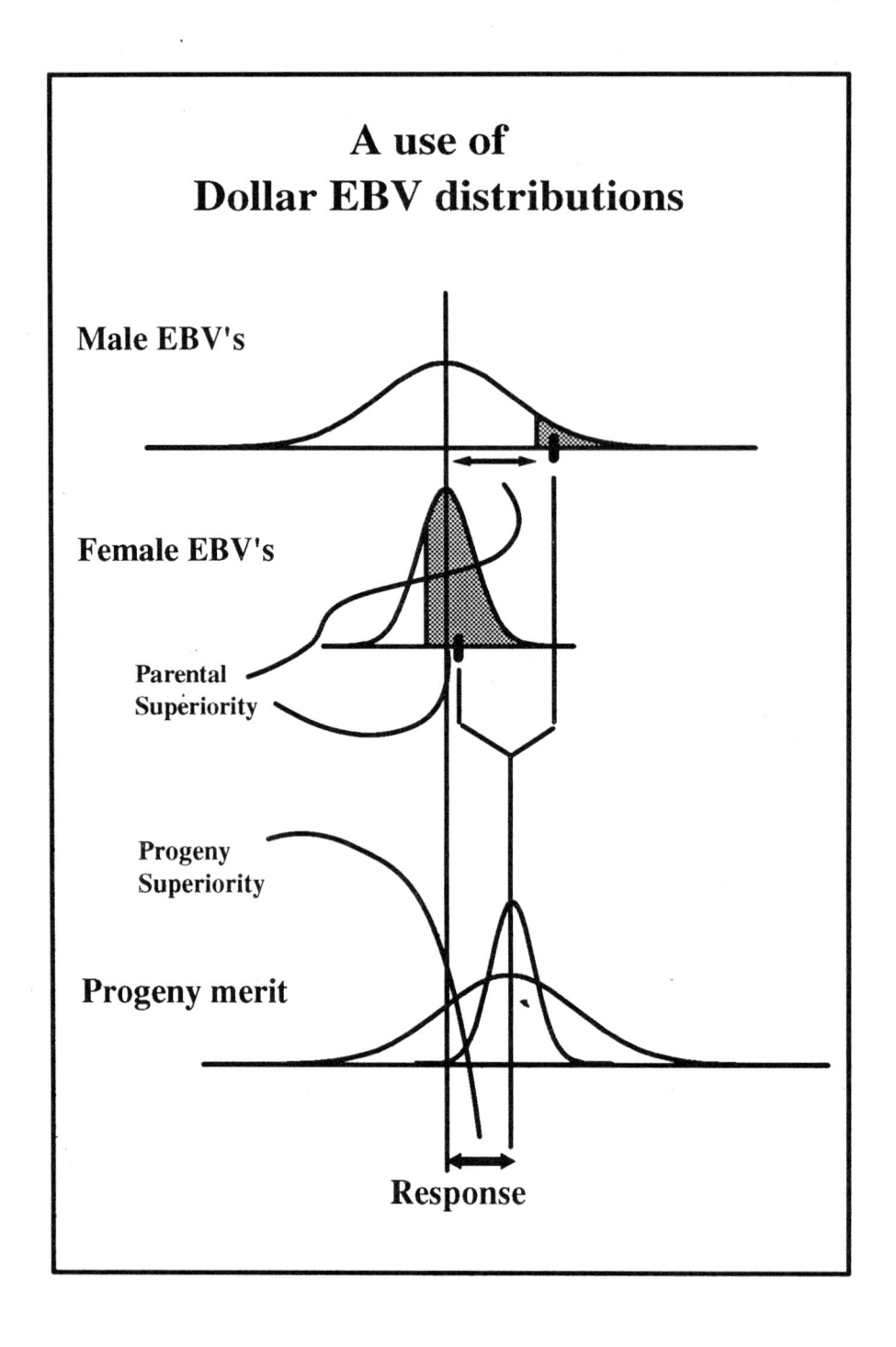

Fig. 4.4 The average EBVs of parents of each sex are simply averaged to predict progeny performance.

Table 4.2 Different strategic objectives require different levels of selection emphasis on direct (EBV$_d$) and maternal (EBV$_m$) sub-traits.

Strategic objective	Select on
Terminal sire: all progeny slaughtered, so	EBV$_d$
Aiming at daughters' calves	$\frac{1}{2}$EBV$_d$ + EBV$_m$
Long term progress: everyone has a mum, so	EBV$_d$ + EBV$_m$

Accuracy of EBVs

It is possible to calculate the standard error (a '±' figure) for an EBV. For example, for EBVs calculated from animals' phenotypes alone, the standard error of these EBVs (SE$_{EBV}$) equals:

$$SE_{EBV} = \sqrt{(1 - h^2) h^2 V_p}$$

- where V$_p$ is the variance of phenotype. The accuracy of EBVs is the correlation between true and estimated breeding values ($r_{A\hat{A}}$), and this can be calculated as:

$$r_{A\hat{A}} = \sqrt{1 - \frac{SE_{EBV}^2}{h^2 V_p}}$$

Breeders have to be careful how they use information about the accuracy of an EBV. An EBV leads to the best prediction of progeny merit, and accuracy cannot be used to improve the correctness of that prediction. Accuracy only indicates the reliability of that prediction. Gamblers might choose animals with less accurate EBVs, as they are just as likely to get more than they bargained for, as they are to get less. On the other hand, a risk averse attitude may lead the breeder to choose animals with more accurate EBVs.

Consideration of accuracy in this way can lead to two problems: Firstly, if the breeder's perspective is narrowed to animals of a certain range of accuracies, there can be missed opportunities to choose animals of high EBV. Secondly, animals with highly accurate EBVs often command a price premium, independently from their EBVs effect on price. The breeder then pays a premium which gives, on average, no extra progeny merit. However, the safety margin obtained from using animals with highly accurate EBVs can be important, depending on the financial profile of the enterprise.

Reference

Taylor & Bogart (1988) Scientific Farm Animal Production, MacMillan

Chapter 5

The Alternative Evaluation Procedures

Markus Schneeberger

Historical Development of Genetic Evaluation Procedures

The development of modern genetic evaluation methodology was largely driven by the population of dairy cows in the United States. The heavy use of artificial insemination made selection of sires the most important and efficient way to obtain genetic progress. Accurate assessment of the genetic merit of sires was important. Thus, most emphasis in early developments was placed on sire evaluations.

The simplest measure to assess the breeding value of an animal is the average performance of its offspring compared to the performance of contemporaries. This simple measure has been used by breeders over thousands of years without knowing anything about Mendelism and statistics.

Daughter-dam comparisons were used to evaluate sires as early as 1900 (according to Engeler, 1957). The advantage of these comparisons is that differential mating of bulls is accounted for, but the problem is the difference in time when dams and daughters perform.

Wright (1932) proposed a bull index (BI) which accounted for the number of dam-daughter pairs:

$$BI = BA + \frac{n}{n+2}(2D - M - BA)$$

where BA is the breed average, D and M are average performance of daughters and dams respectively, and n is the number of daughter-dam pairs.

Lush (1933) introduced the heritability into these formulae, but the problem of the difference in time when dams and daughters perform remains. In later developments, this problem was avoided by using deviations from herdmate performances.

Hazel's selection index (Hazel 1943) laid the basis for the developments of numerous genetic evaluation systems for several species some of which are still in use and are still being developed today.

The contemporary comparison of Robertson and Rendel (1954) used a formula of the following basic skeleton to calculate an Estimated Breeding Value (EBV) of a sire:

$$EBV = 2 \cdot b \cdot (\bar{D} - \bar{C}),$$

where

$\bar{D}$ = average performance of daughters

$\bar{C}$ = average performance of contemporaries, in most applications cows of similar age, calving in the same herd-year-season

b = regression adjusting the deviation of daughter to contemporary performance according to the genetic information content, i.e., accounting for the heritability of the trait and the distribution of daughters and contemporaries.

In this method it is assumed that the environmental herd-year-season effects are known, and that all bulls are mated to representative samples of cows.

The herdmate comparison method allowed for differences in genetic merit between herds by the assumption that part of the variation of the mean herd performances was due to genetic differences between herds.

The Modified Contemporary Comparison (Dickinson *et al.* 1976), used by the United States Department of Agriculture until 1989 to predict breeding values for dairy bulls and cows, used an iterative approach to account for differences in genetic merit of mates and between herds.

Best Linear Unbiased Prediction (BLUP)

Best Linear Unbiased Prediction (BLUP) to predict breeding values was described by Henderson (1973). The effects of environment and the breeding values of animals are estimated simultaneously and, therefore, genetic differences between herds are correctly accounted for. The first applications used a sire model which still made the assumption that all sires are mated to cows of equal genetic merit. Later, sire-maternal grand sire models partly relaxed this assumption.

BLUP animal models are now used in many countries for a number of species including dairy and beef cattle, swine, horse, sheep and fish. They incorporate all relationships among the animals and predict breeding values for all animals. Genetic differences between mates of different sires are therefore correctly account for.

In applications where a series of traits is recorded over time, e.g. weights at different ages, and selection is practised on these records, an analysis which deals with each trait separately produces biased EBVs. This can be overcome by analysing all traits simultaneously in a multiple-trait model where the genetic relationships (genetic correlations) among the traits are accounted for. This method is used, e.g. in the BREEDPLAN International genetic evaluation system for beef cattle.

Statistical method

The central problem in predicting breeding values from observed phenotypic values is to separate genetic from environmental effects. In statistical terms the problem is to simultaneously estimate constants for fixed (environmental) effects and predict realised values of a random variable (breeding values of individual animals). The solution to this problem is to obtain Best Linear Unbiased Estimates (BLUEs) for the fixed effects and Best Linear Unbiased Predictions (BLUPs) for the realised values of the random variable.

> Best: Estimates and predictions have minimum variance
> Linear: Solutions are a linear combination of the observations
> Unbiased: Expectations of solutions are equal to expectations of true values

The equation for an observation can be written as:

$$y_{ij} = F_i + u_j + e_{ij}, \tag{1}$$

where
- y_{ij} = observation in i^{th} level of fixed effect and j^{th} level of random effect
- F_i = i^{th} level of fixed effect
- u_j = j^{th} level of random effect
- e_{ij} = random residual effect associated with this observation

u_j and e_{ij} are randomly distributed variables, and assumptions have to be made about the distribution.

The set of equations (1) for all observations in a data set can be expressed in matrix notation:

$$y = Xb + Zu + e, \tag{2}$$

where
- y = vector of observations
- b = vector of fixed effects
- u = vector of random effects
- e = vector of random residuals
- X = design matrix connecting the observations to the fixed effects
- Z = design matrix connecting the observations to the random effects

Best Linear Unbiased Prediction (BLUP) has proved to be a very useful method to predict breeding values of animals (realised values of a random variable) while simultaneously taking into account systematic environmental effects (estimating constants for fixed effects).

Classification of models

The models most commonly used for the prediction of breeding values of animals can be classified in two different ways:

- Classification of models according to the definition of the random effect:

    Sire model. The random effects are the effects of the sires of the observed animals, i.e., half the sires' breeding values. In most applications, a single fixed effect is used to account for differences in environment where the animals perform, e.g. herd-year-season effects. The equation for an observation, e.g. a lactation record of a cow becomes:

    $$y_{ijk} = H_i + s_j + e_{ijk},$$

    where
    - y_{ijk} = observation on animal ijk, sired by sire j, in the ith level of the environmental effect
    - H_i = fixed effect of the ith level of environment (herd-year season)
    - s_j = effect of the jth sire
    - e_{ijk} = random residual effect associated with this observation

    s_j is randomly distributed with mean 0 and variance σ_s^2 which is equal to 1/4 of the additive genetic variance σ_a^2.

    e_{ij} is randomly distributed with mean 0 and variance σ_e^2.

    In matrix notation the equations become:

    $$y = Xb + Zs + e,$$

    where
    - y = vector of all observations
    - b = vector of fixed effects
    - s = vector of random sire effects
    - e = vector of random residuals

    X and **Z** are design matrix connecting the observations to the fixed and random effects, respectively

    The vector of sire effects has an expectation of zero and variance $\mathbf{Var(s)} = \mathbf{I}\sigma_s^2$ if the sires are unrelated, **I** is an identity matrix. If the sires are related covariances exist among them, and $\mathbf{Var(s)} = \mathbf{A}\sigma_s^2$, where **A** is the Numerator Relationship Matrix (NRM) consisting of the additive genetic relationships among the sires. The distribution of the random residuals is characterised by an expectation of zero and variance $\mathbf{Var(e)} = \mathbf{I}\sigma_e^2$ which implies the assumption of uncorrelated residuals.

    Sire - maternal grandsire model. This model is an extension to the sire model. It connects an observation through the **Z** matrix not only to the effect of the sire of the animal but also to half the effect of the maternal grandsire.

Animal model. Breeding values for all animals are predicted in an animal model. The observation on an animal is described by the following equation:

$$y_{ij} = H_i + a_j + e_{ij},$$

where

H_i = effect of the ith level of environment

a_j = breeding value of animal j, distributed with mean 0 and variance σ_a^2, the genetic variance for the observed trait, and

e_{ij} = random residual effect associated with the observation yij, distributed with mean 0 and variance σ_e^2.

The equations for all observations can also be written in matrix notation:

$$y = Xb + Za + e,$$

where

a = vector of the breeding values of the animals, and all other symbols are as defined above

The assumptions about the variances of the random effects are:

Var(a) = **A** σ_a^2, where **A** is the numerator relationship matrix (NRM)

Var(e) = **I** σ_e^2, where **I** is an identity matrix.

The use of the numerator relationship matrix allows the inclusion of breeding values of animals without observations, e.g. sires, into the breeding value vector **a**. The animal model has become the method of choice and will be dealt with further in a special section below.

Classification of models according to the treatment of traits:

Single-trait model. Only one trait is analysed.

Multiple-trait model. More than one trait is analysed simultaneously, taking into account the relationships (genetic and environmental correlations) among the traits.
The vector **y** of equations (2) contains observations on up to **m** traits per animal, and the vector **u** consists of breeding value for **m** traits per animal rather than only one, where **m** is the total number of traits in the analysis, some of which might not be observed.

In the case of the animal model, the variance of the breeding values of the animals, **a**, now becomes

$$\text{Var}(\mathbf{a}) = \mathbf{G} = \mathbf{A} * \mathbf{G_0},$$

where **A** is the NRM as before, and $\mathbf{G_0}$ is the matrix of genetic variances and covariances among the m traits.

The symbol $*$ denotes the direct or Kronecker product.

Likewise, **Var(e)** takes on the more general form **R** which is block diagonal with the diagonal blocks consisting of the residual variances and covariances among the traits observed for each animal.

Models with repeated measures. More than one measurement of the same trait is analysed, in a single-trait or a multiple-trait model.

The numerator relationship matrix

The Numerator Relationship Matrix (NRM) plays a central role in genetic evaluation using BLUP, particularly when an animal model is used. Its diagonal elements are one plus the inbreeding coefficient of the animal, and the off-diagonal elements are the additive genetic relationships among the animals, e.g. half for a parent-offspring or a full sib relationship, one quarter for the relationship between grand parents and grand progeny or between half sibs, etc.

Consider the following pedigree structure:

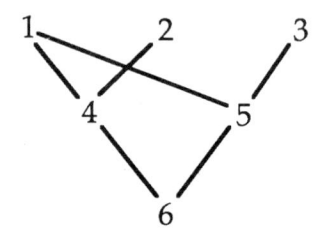

Animal 4 is the offspring of animals 1 and 2, animal 5 has parents 1 and 3, and 6 is the offspring of 4 and 5 who are half sibs. Therefore, animal 6 is inbred. This affects not only the diagonal element of animal 6 but also the relationship coefficients with its parents and the common ancestor 1 are increased. The NRM for these animals is:

$$\begin{array}{c c} & \begin{array}{cccccc} 1 & 2 & 3 & 4 & 5 & 6 \end{array} \\ \begin{array}{c} 1 \\ 2 \\ 3 \\ 4 \\ 5 \\ 6 \end{array} & \left[\begin{array}{cccccc} 1.0 & 0 & 0 & 0.5 & 0.5 & 0.5 \\ 0 & 1.0 & 0 & 0.5 & 0 & 0.25 \\ 0 & 0 & 1.0 & 0 & 0.5 & 0.25 \\ 0.5 & 0.5 & 0 & 1.0 & 0.25 & 0.625 \\ 0.5 & 0 & 0.5 & 0.25 & 1.0 & 0.625 \\ 0.5 & 0.25 & 0.25 & 0.625 & 0.625 & 1.125 \end{array} \right] \end{array}$$

As will be seen in the next section, the inverse NRM is needed to compute estimated breeding values for the animals. Fortunately, the inverse NRM has a very simple structure and simple rules (Henderson 1976) can be followed to construct it directly from a list of the animals, their parents and inbreeding coefficients. Methods for rapid computation of inbreeding coefficients are available for large population sizes (Tier 1990).

Obtaining the Solutions

Let's consider the multiple-trait animal model:

$$y = Xb + Za + e, \text{ with}$$

$$Var(a) = G = A * G_0, \text{ and}$$

$$Var(e) = R.$$

The individual symbols are as defined previously.

Solutions for the fixed effects **b** and the random breeding values **a** are obtained by solving the following set of equations, called Henderson's Mixed Model Equations (MME), where

$\hat{b}$ is the vector of solutions for the fixed effects b and

$\hat{a}$ is the vector of solutions for the random breeding values a.

$$\begin{bmatrix} X'R^{-1}X & X'R^{-1}Z \\ Z'R^{-1}X & Z'R^{-1}Z + G^{-1} \end{bmatrix} \begin{bmatrix} \hat{b} \\ \hat{a} \end{bmatrix} = \begin{bmatrix} X'R^{-1}y \\ Z'R^{-1}y \end{bmatrix}$$

$$G^{-1} = A^{-1} * G_0^{-1}$$

$$\begin{bmatrix} \hat{b} \\ \hat{a} \end{bmatrix} = \begin{bmatrix} X'R^{-1}X & X'R^{-1}Z \\ Z'R^{-1}X & Z'R^{-1}Z + G^{-1} \end{bmatrix}^{-} \begin{bmatrix} X'R^{-1}y \\ Z'R^{-1}y \end{bmatrix}$$

$^{-}$ denotes the generalised inverse operation.

In most practical applications it is not feasible to invert the coefficient matrix of the MME. Other methods to solve these equations are therefore employed, such as Gauss-Seidel or Jacobi iteration.

The coefficient matrix of the MME is very sparse i.e. contains many zeros, and consists of multiple repetitions of a small number of submatrices. It is therefore not necessary to store the complete matrix. A list of animals with their parents and inbreeding coefficients and an indicator on what traits have been observed in what level of the fixed effect, together with one copy of each submatrix are enough to represent the complete set of equations. This way of representing the MME is known as the Implicit Animal Model (Tier and Graser 1991) and a considerable amount of computer storage and computing time can be saved. This makes the application of multiple-trait animal model BLUP for the prediction of breeding values feasible even for very large populations.

Accuracy of estimated breeding values

EBVs can be estimated from vastly different amounts of information, e.g. from a single observation on a distant relative, or a number of observations on the animal itself and on many progeny. The information a particular observation contributes to an EBV depends also on the number of animals which the observed animal is compared with and its genetic relationship to these animals. The accuracy of an EBV is expressed as the correlation between the true and the estimated breeding value, or as a function of this correlation. In the BREEDPLAN International genetic evaluation system, this correlation is expressed as a percentage. Many systems particularly in Europe, and for US dairy cattle report the square of this correlation, and a transformation of the correlation to a linear scale, the so-called BIF Accuracy is used in the US for genetic evaluation of beef cattle.

The BLUP Animal Model

This section is devoted to the BLUP animal model to stress its importance to modern genetic evaluation systems. It employs a simple concept which, at the same time, makes it the most powerful tool to predict breeding values for animals currently available.

Parents produce gametes which contain a sample half of their genes. The gametes of the two parents combine randomly to generate the offspring. The breeding value (BV) of an offspring can then be expressed with the following simple formula:

$$BV_{offspring} = \frac{1}{2} \{ BV_{sire} + BV_{dam} \} + \text{Mendelian sampling}$$

An estimate of the breeding value of an animal can be obtained by adding half the estimated breeding value of the parents plus an estimate for the Mendelian sampling. This last term which accounts for half of the genetic differences between animals can only be estimated from records of the animal itself and from the estimated breeding values of its progeny. The estimated breeding values of the parents and progeny, in turn, contain information stemming from their parents and progeny. Thus, the information from all relatives is incorporated into each animal's Estimated Breeding Value.

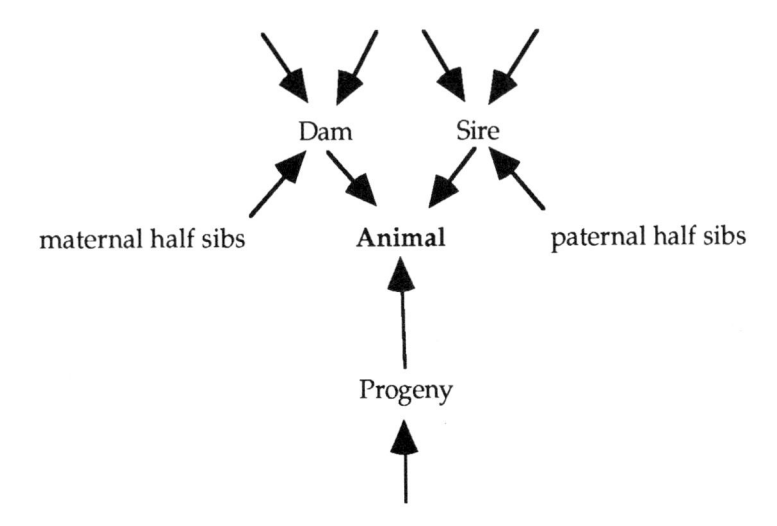

```
              ↓↙  ↓↙
              Dam    Sire
             ↗↘    ↗↘
maternal half sibs   Animal   paternal half sibs
                       ↑
                    Progeny
                       ↑
```

The BLUP method allows the simultaneous computation of EBVs for all animals while at the same time accounting for differences in environment under which different animals perform.

As shown previously, the models can be classified according to the number and treatment of traits analysed:

Single trait, one measurement. This is the most simple case. Examples would be the prediction of breeding values for first lactation milk yield or birth weight which are only observed once on a given animal.

Single trait, repeated measurements. This case is used for situations where a trait is observed more than once on a given animal; e.g. milk yield observed in several lactations, scrotal size of a bull measured at different ages, conception rate of a female observed in different parities. Only one EBV, however, is computed per animal, e.g. an EBV for milk yield, scrotal size or conception rate. The use of more than one observation increases the accuracy of the EBVs. The genetic evaluation for dairy cows in the United States and in some European countries uses a single-trait animal model with repeated records.

Multiple traits, one or repeated measurements per trait. The genetic relationships (correlations) among traits are exploited in these types of models. Observations on genetically correlated traits add information to the estimation of the breeding values for the other traits and, therefore, increase the accuracy of the EBVs. Examples of correlated traits are scrotal size measured in males and conception rate measured in females, or weights taken at 200, 400 and 600 days of age. Subsequent lactation records of a dairy cow can also be regarded as correlated traits rather than as repeated records of the same trait. Multiple-trait animal models are used, e.g. for the genetic evaluation of dairy cows in Germany, three parts of the first lactation and the yields in second and third lactation are treated as correlated traits, and for genetic evaluation of beef cattle in BREEDPLAN International where repeated measurements are also accommodated.

Maternal effects. Maternal effects are a special case of a multiple-trait analysis. A measurement taken on an animal is not only an observation for its own direct genetic effect but also for its dam's maternal genetic effect. The classical example is the weaning weight of a calf which is an expression of the calf's genetic potential to grow and, at the same time, of the dam's genetic potential to provide milk for the calf. Figure 5.1 provides an illustration for these genetic pathways. The BREEDPLAN International genetic evaluation system estimates breeding values for maternal effects for birth and 200-day (weaning) weight.

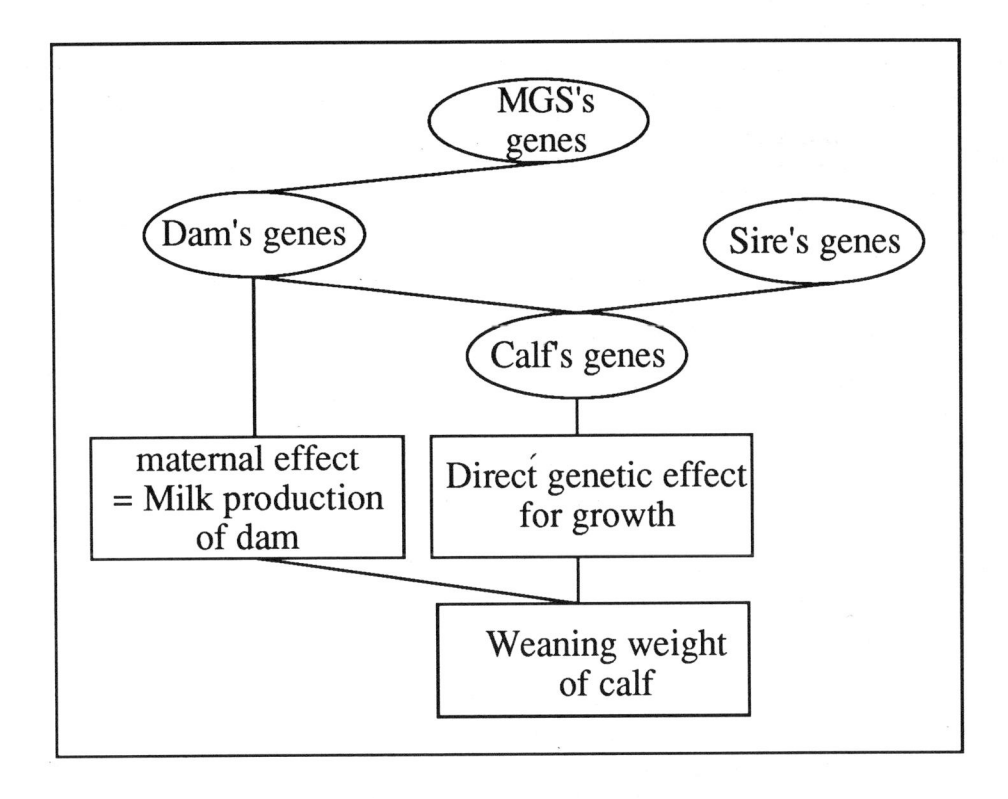

Fig. 5.1 Genetic pathways for an observation of weaning weight as an example for a trait with a maternal effect (MGS = maternal grand sire)

Genetic Evaluation for Categorical Traits

Some traits are observed subjectively in distinct categories rather than on a continuous scale. Examples are:

- Survival to a certain age - the animal either has survived (category 1) or not (category 0)
- Stillbirth - the animal is either born alive (category 1) or not (category 0)
- Calving ease - this trait is scored in several categories. In Australian beef cattle for example, these categories are defined as,

    1 No assistance   4 Surgical delivery
    2 Easy pull   5 Malpresentation
    3 Difficult pull

The first two examples, survival and stillbirth, are 'all-or-none' categories, i.e., an event is either observed or not. Thus, there are two categories involved. The above calving ease scoring system is an example for more than two categories.

The BLUP method used for genetic evaluation of continuously distributed traits, such as weight or milk yield are no longer valid for genetic evaluation of categorical traits,. although, they may yield satisfactory results in practice in some circumstances, especially when more than about 4 categories are involved.

In the analysis of categorical data it is normally assumed that the categorical trait has an underlying continuous distribution (e.g. Falconer 1981, chapter 18). This concept is illustrated in Figure 5.2 for the case with two categories.

The underlying variable may be thought of as the concentration of some substance. When the concentration is above a certain value, called the threshold **T**, a certain characteristic e.g. a stillborn calf is observed. When the concentration is below the threshold, this characteristic is not observed e.g. a live calf is born. The actual value of the underlying variable cannot be observed. For example, the difference between a stillborn calf with the value of the underlying variable being close to the threshold and another stillborn calf with the value of the underlying variable being on the far end of the distribution cannot be detected.

In groups of animals, such as members of a herd or progeny of a sire, the percentage or incidence of an observed category varies across the range of the scale of the underlying variable. From the incidence, the mean for that group of animals can be estimated and the difference between this and another group of animals can be assessed.

The concept of underlying variables, as illustrated in Figure 5.2, can be extended to the case of more than two categories with more than one threshold.

A method for the prediction of breeding values for categorical traits using the concept of the underlying variable has been described by Gianola and Foulley (1983). Breeding values are predicted on the underlying continuously distributed variable. The predicted breeding values can be transformed back to the categorical scale and incidences of specific categories can be given for various combinations of fixed effects, e.g. percentage of stillborn calves from first-calf heifers calving in spring.

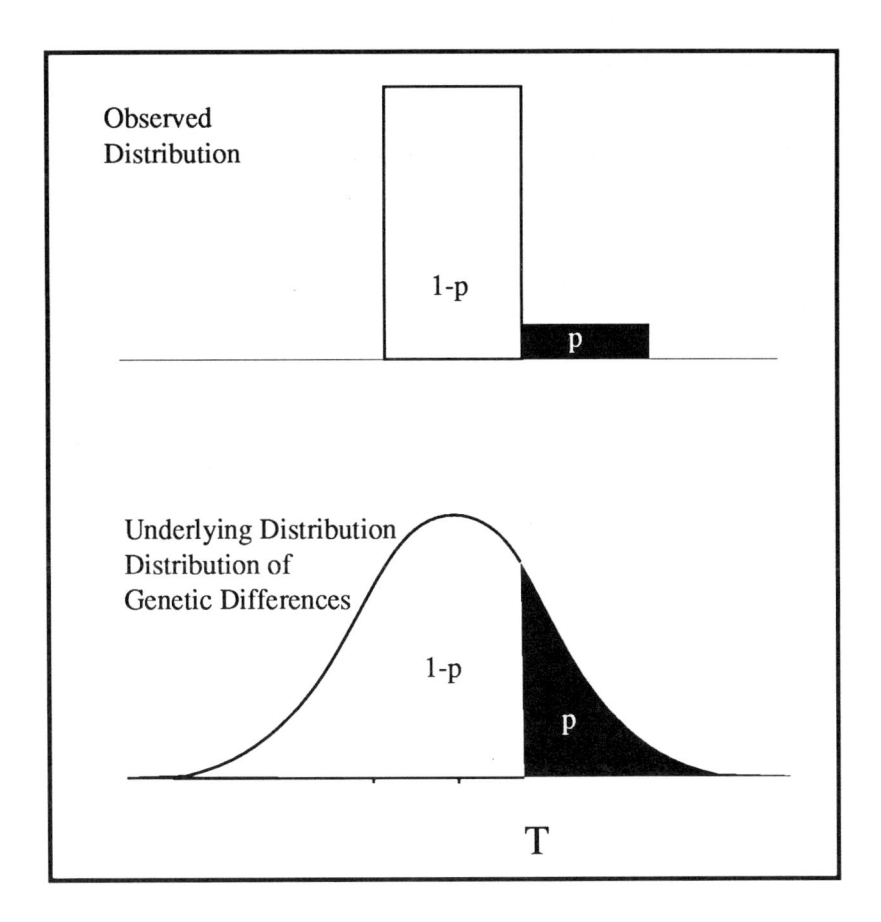

Fig. 5.2 Observable discontinuous and non-observable, underlying continuous distributions with threshold T and proportion p of animals expressing the characteristic.

References

Dickinson FN, Powell RL, Norman HD (1976) An introduction to the USDA-DHIA Modified Contemporary Comparison. In: The USDA-DHIA Modified Contemporary Comparison Sire summary and cow index procedures. Agric. Res. Service. USDA Production Res Report No. 165: 1

Engeler W (1957) Die Nachkommenprüfung auf Milch- und Fleischleistung beim Rind. Schriften Schweiz Vereinigung für Tierzucht Nr. 22

Falconer DS (1981) Introduction to quantitative genetics. Longman, London and New York

Gianola D, Foulley JL (1983) Sire evaluation for ordered categorical data with a threshold model. Genet Sel Evol 15: 201-223

Hazel LN (1943) The genetic basis for constructing selection indexes. Genetics 28: 476-490

Henderson CR (1973) Sire evaluation and genetic trends. Proc Anim Breed Genet Symp in Honor of Dr. J.L. Lush. ASAS and ADSA Champaign, Illinois pp 10-41

Henderson CR (1976) A simple method for computing the inverse of a numerator relationship matrix used in prediction of breeding values. Biometrics 32: 69-83

Lush JL (1933) The bull index problem in the light of modern genetics. J Dairy Sci 16: 501-

Robertson A, Rendel JM (1954) The performance of heifers got by artificial insemination. J Agr Sci 44: 184-192

Tier B (1990) Computing inbreeding coefficients quickly. Genet Sel Evol 22: 419-430

Tier B, Graser H-U (1991) Predicting breeding values using an implicit representation of the mixed model equations for multiple trait animal model. J Anim Breed Genet 108:81-88

Wright S (1932) On the evaluation of dairy sires. Proc Am Soc Anim Prod: 71

Chapter 6

Within Versus Across Herd or Flock Evaluation

Keith Hammond

Chapter 4 described the principles of modern estimation of genetic merit, the EBV or EPD and Genetic Trends, and chapter 5 outlined the historical development and current methodology. In this chapter we describe the field procedures used to obtain valid genetic evaluations across a number of herds or flocks.

Across-Herd or Flock Comparisons are Important?

Maximising gains from breeding involves locating and using the best available genetic material. The seedstock producer wishes to remain competitive, so must locate and use the best, and the producer of commercial product is continually being pressured to increase productivity and change product quality, so commercial producers must also locate and utilise the best genetic material for the money available. How much should the commercial producer spend on superior genetics for a particular production-marketing strategy, and how is this calculated?

Rarely then will the **seedstock producer** be satisfied with utilising only genetic material from within the herd or flock, for:

- Other herds or flocks may be substantially superior on average genetic merit for the traits of interest; and
- Being able to select from many herds or flocks i.e. a much larger number of animals than in their own, increases the probability of finding an exceptional animal.

The **commercial producer** buying bulls, rams or boars needs to know the location of the best available genetic material that money can buy.

Hence, both seedstock producers and the buyers of seedstock will be interested in comparing animals on genetic merit across herds or flocks. The difficulty in doing this is that virtually all traits of interest are imperfectly inherited. Management and other differences between herds or flocks will mean that most selection decisions based on direct comparisons of animals across herds or flocks, whether visual or otherwise measured, will be wrong, i.e. are unlikely to identify the animals which are genetically the best.

Across-Herd or Flock Procedures

How then to make valid across-herd comparisons? The problem boils down to devising ways of removing the non-genetic biases, such as management differences, and of utilising as much information as possible, e.g. on relatives, to obtain reliable or accurate rankings on genetic merit across the herds or flocks involved, for the traits of interest. There are various field strategies to achieve this and their value varies between:

- Species;
- Industry structures;
- Availability and cost of artificial breeding technologies;
- Traits of interest; and
- Organising ability of the people involved.

Further, in some situations two or more of the available strategies will be used.

What are these field procedures? Basically, they involve either removing the animals to be compared, or their sibs or progeny to a central location (Central Testing), or producing sufficient progeny from one or more common sires in each of the herds or flocks across which comparisons are to be made. Only then will valid comparisons of all animals of interest be practicable following suitable analysis to remove the between-herd/flock non-genetic biases from each animals record for each measurement.

Where sufficient AI is used by the herds or flocks across which comparisons are to be made, these AI sires are used **across** these herds or flocks, **and** each AI sire has the same identification wherever it is used, then the across-herd analysis for genetic merit is little more than a large within-herd analysis. Where this is not so then the necessary linkages must be created by either introducing Reference Sires through AI or via the Ram Circle concept. The latter was first developed in Norway and rams are sequentially moved between the flocks at mating across which comparisons are to be made.

It is important to note that neither the Central Test operation nor the widespread use of AI, nor the Reference Sire or Ram Circle scheme are genetic improvement or breeding programs in themselves. This is only so when the best animals from the regular evaluation analysis are selected and used back in the same population of herds or flocks that generated them, with this process continuing over time. If for example, one group of say, registered herds or flocks contribute the animals for central testing, and the best of these animals are subsequently used but only in the commercial industry, and no other genetic gains are being achieved in the registered herds or flocks, then virtually no continuing gains will be achieved in the commercial industry. There are many examples in many countries where this interrupted flow of genes has occurred.

There is another across-herd/flock procedure which can facilitate across-herd/flock comparisons, the Open Nucleus Breeding Scheme. Here, a number of herds or flocks contribute their best females to a central nucleus herd or flock and in return each year receive the males they need for replacements from the nucleus. This across-herd procedure **is** a breeding program and not simply a genetic evaluation procedure.

Table 6.1 summarises some of the primary characteristics of central testing, marker or reference sire procedures and nucleus breeding programs.

Table 6.1 Types of across-herd or flock operations associated with genetic evaluation, primary reasons for their formation and their problems

Across Herd/Flock Strategy	Why?	Problems
Central Testing - Progeny - Sib - Clone - Performance	- Locate/test males over (small) herds/flocks - Measure difficult traits (intake, carcase quality) - Sell stock - Compare herds/flocks	- Pre-test environment and repeatability - Cost/Organisational - Sampling animals for Testing - Only some measurements on some stock
Marker (Reference) Sire (and Ram Circles)	- Increase N^* - Across herd/flock genetic evaluation - Locate AB sires - Accreditation schemes	- Progeny per marker per herd/flock - Sire x Herd/Flock effects - Organisational
Nucleus Program	- Increase N - Reduce costs of recording and of replacements - Increase A^* - Large S* in establishment ←	- Mainly organisational - Increase in genetic gain not great from theory - G x E Often not realised

*N is total population size, A is Accuracy of selection and S is selection differential

Central Testing

The difficulty with central testing is that for traits such as growth and milk production, the carry-over or pre-test environmental effects can remain for some time so these continue to bias the comparisons between animals during central testing. In addition, traits such as reproductive performance of females generally can not be measured in the central test as usually only young animals are evaluated centrally for traits such as growth. Hence, most modern breeding operations endeavour to combine on-farm performance recording with central testing. Central testing has the advantage of centrally locating costly equipment, so enabling the measurement of traits such as feed intake which otherwise could not be measured back on-farm.

Of course, wether trials are a form of central testing where they are used to evaluate sires. Feedlots may form a central test location where sire progeny groups are identified from the producing herds through the lot whilst measurements are being taken and even into the abattoir for carcase and meat quality evaluation. However, logical difficulties remain for seedstock herds or flocks attempting to do this - see Sections 3.5 and 6.6.

Sire Referencing

Experience shows that reference sire operations can generate substantial benefits if properly applied and conducted. They offer:

- Improved ability for buyers of seedstock to discriminate between herds or flocks.
- The chance for seedstock producers to increase returns from males identified in the reference sire system as superior, and from the herd or flock as a whole.
- Reliable evaluation for the less heritable traits - at least for the reference sires for which the total number of progeny across all herds or flocks will be large, generating reliable genetic evaluations for traits such as ease of calving, provided a sound scoring system for calving ease is used across all herds.
- A means of introducing new variation to the herd or flock.
- A supply of screened males for artificial breeding centres, with an increased number of progeny from the best.
- Increased genetic gain in the longer term, providing the best of the tested sires are selected and used as the next crop of reference sires.
- A powerful data base for the group of participants and the breed for research and promotion.
- An educational tool for promoting performance recording in genetic evaluation.
- Discount semen prices for reference sires where these prices are used by the organising agency as an incentive to increase involvement and to help prevent introduction of bias from breeders preferentially treating the progeny of high priced sires.

Organisation of the field operations is important if the information content is to be maximised in the resulting data sets. For example:

- Unique, permanent and accurately recorded identification of all offspring and their parents is necessary.
- A representative group of females should be joined to each reference and home sire used, particularly in the early years of a scheme, even where BLUP procedures are used for the subsequent evaluation analyses.
- Care is needed with the use of clean-up (natural mating) sires - use either differential colour marking or a lapse of at least three weeks in cattle between AI and the introduction of clean-up bulls.
- Female mating groups should be rearranged each year to improve linkage.
- Management groups, groups of animals treated differently at any point in time, should be arranged to ensure that all are strongly linked within the herd or flock and contain progeny of as many sires as possible, with a minimum of two sires being represented in each management group.

- All management groups should be coded and recorded, otherwise non-genetic biases will remain in EBVs after analyses are completed.
- Selectively culling some offspring prior to their first measurement will introduce biases for traits such as growth rate.
- All dams should be recorded where genetic evaluation for traits such as female reproductive performance is intended sooner or later.
- The best data set will be formed when the herds or flocks involved remain in the system over time - the greater the herd or flock turnover the lower the information content of the data set as a whole.
- It is advisable for herds or flocks to be familiar with the use of Artificial Insemination and of Performance Recording prior to becoming involved in across-herd/flock schemes.

Across-Herd or Flock Analyses

A central agency. The agency or bureau performing the regular analysis for the across-herd/flock evaluation will introduce further checks prior to analysing the data to ensure that all herds/sire groups are comparable - the BLUP theory actually allows the calculation of EBVs for animals which are not involved in direct or indirect comparisons for one or more correlated traits. However, seedstock producers may not readily accept such EBVs as valid comparisons, particularly when they are encouraged to maximise the number of comparisons within and between herds for the most reliable analysis.

Genotype by environment interactions. The importance of genotype-environment interactions is frequently raised, particularly where participant herds or flocks are widely distributed and production systems/markets vary. Whilst, these may not be a particularly serious problem in intensive pigmeat production, there will almost certainly be substantial G by Es for at least some traits within breeds in the beef and sheep industries. However, reliably describing G by Es requires a large data set with sires being well used across the range of environments. Currently there are few such data sets. Once reliable estimates of G by Es or, alternatively, genetic correlations between major environments, e.g. grass-finishing *versus* lot finishing of beef, are available then several different possibilities exist for accommodating these in breeding operations, from conducting separate EBV analyses for the different environments to providing for the G by Es in the one analysis across all environments.

Combining On-Farm with Central Test data. Modern genetic evaluation services for some industries will eventually need to combine on-farm data from many herds or flocks with central test station data. To do this successfully the whole data set must be correctly structured and linked to obtain valid comparisons of all animals.

Interim EBVs or EPDs. Genetic evaluation systems have now developed to the point where a herd or flock involved in the across-herd analyses, which are generally conducted at regular intervals during the year when sufficient additional data has accumulated and before the next mating season, can obtain an interim analysis immediately it submits its next set of records. The resulting interim EBVs will combine both this latest data from within the herd or flock and the results from the previous across-herd analysis.

Comparing Genetic Trends. Modern across-herd genetic evaluations enable the individual participants to compare their rates of genetic change with that for all herds involved in the analysis, for each trait analysed. Again, this provides additional information for the decision processes in breeding.

Across-Herd or Flock Analyses for Meat Yield and Quality

For traits such as meat quality in cattle, across-herd genetic evaluation becomes quite difficult logistically, primarily because seedstock producing herds slaughter very few animals. The seedstock producers then need to arrange for sires to also be mated in commercial herds to produce sufficient slaughter progeny and reliable genetic comparisons for these sires. Of course, in future when multiple cloning of embryos becomes practical, then just four to six clones of a male embryo may be grown out to slaughter for genetic evaluation.

Again, the superior alternative in at least some situations will be the development of indirect measures of carcase and meat quality on the live animal. Providing these measures are reasonably heritable and have high genetic associations with the actual quality traits of interest, the seedstock producing herds will then be able to measure all animals, to create much more information on their relative genetic value than would be obtained through the costly and time consuming progeny testing of a small number of their sires via commercial slaughter progeny.

Chapter 7

Genetic Evaluation of Beef Cattle

Alex McDonald

Introduction

Objective measurement and genetic evaluation of beef cattle in Australia developed very slowly during the 1970s and early 1980s. Prior to 1972 the concept of performance recording was promoted by individual state Departments of Agriculture. Adoption was limited to a few innovative stud and commercial breeders. In 1972 the National Beef Recording Scheme (NBRS) was formed and an Adjusted Weight Ratio analysis was provided for individual herds by the Agricultural Business Research Institute (ABRI) at the University of New England.

Individual weight records were simply adjusted for age of the calf, sex of the calf and age of the dam and compared with contemporary groups, i.e. calves born within 60 days and run in the same paddock. The adjusted weights were expressed as a percentage of the average adjusted weight of the group. This was a phenotypic measure not a genetic measure. Comparisons were limited to animals reared in the same paddock in the same herd and born within a 60 day calving span.

The first genetic measures of performance introduced to the industry were across-herd evaluations done for two European breeds in the late 70s. The amount of AI used to introduce these breeds into Australia provided very strong linkage between the herds involved and a **Regressed Contemporary Comparison** procedure similar to that then being used in the dairy industry were applied to the sire contemporary group means after adjusting all weight records for calf age and dam age. The regression factor was determined by the heritability of the trait and the number of sire progeny and of contemporaries within each contemporary group. Following this, the regressed differences were combined across all groups in which each sire was represented.

The more balanced the data, the better the regressed contemporary comparison procedure works, but it assumed that preferential mating, prior culling and even genetic progress itself were not occurring. Equal sire representation in all contemporary groups is best.

The industry was generally unimpressed. By 1984 only about 200 breeders from across Australia were actively performance recording with the National Beef Recording Service and only the Simmental and Limousin breeds had attempted a genetic evaluation using the Regressed Contemporary Comparison analysis for their breed.

The Animal Genetics and Breeding Unit (AGBU) recognised the potential of Best Linear Unbiased Prediction (BLUP) theory to overcome many of the shortcomings of the earlier procedures. BLUP technology was first used for an across herd analysis for the Simmental breed in 1982 employing a multitrait sire maternal grandsire model. This early learning step led to the introduction of the multiple trait animal model of BLUP in 1985. The system was initially available for within herd evaluation of growth traits under the banner of

BREEDPLAN. The first GROUP BREEDPLAN or breed analysis was conducted for the Angus breed in 1986 involving just 14 herds.

Since that time the number of Australian herds using the BREEDPLAN system has increased to 1300 and GROUP BREEDPLAN analysis are conducted annually for 8 breeds. In addition the GROUP BREEDPLAN technology has been licensed to Breed Associations in the USA and New Zealand.

The attributes of the multiple trait animal model of BLUP used in BREEDPLAN and GROUP BREEDPLAN are:

- **EBVs for all animals.** BREEDPLAN simultaneously estimates breeding values for all animals and all traits in the analysis.
- **All information.** It uses all information from each animal, from all their past and present relatives, and from the contemporaries they are compared with when calculating the EBVs, and finally, it also utilises the information from genetically correlated traits.
- **Removes non-genetic biases.** Adjustments are done in ways which remove the majority of the major biases in records due to such factors as paddock and other treatment differences, calf age and dam age effects and sex biases.
- **Removes genetic biases.** Account is taken of effects such as preferential mating, unequal competition between contemporary groups, prior culling and the genetic differences between age groups resulting from selection itself.
- **Produces genetic trends.** Each round of EBV reports also includes an update of the genetic change being achieved in each trait measured. Only the more advanced BLUP procedures are capable of this. It serves as a guide to the effectiveness of the total breeding operation.
- **Promotes effective breeding practices.** Using all past and present information emphasises the need for correct pedigrees, measurement and recording. The emphasis on group size increases selection accuracy and the potential for selection. In addition, clear signals are provided when current sires and dams are no longer superior and need replacing.

Changing Attitude of Breed Associations

When the BREEDPLAN and GROUP BREEDPLAN systems first became available most of the herds were recording performance data independently of breed associations pedigree records. The Simmental and Limousin breeds were the only breeds where breeders were actively recording performance data on the breed association data base. This is the preferred method of recording performance because the BLUP System is able to make maximum use of the pedigree information available on the breed association data base.

There has been a major change in attitude by Australian breed associations over the last 5 years. Most of the breed associations which run GROUP BREEDPLAN evaluations now provide the facility and encourage members to measure performance and record it on the breed association data file. They have encouraged their members to transfer their accumulated performance records onto the breed association data base, up to 20 years of records for some herds.

Secondly, many breed associations now allow unregistered animals to be recorded on the breed society data base at a considerably lower fee than for registered animals. This allows both registered and unregistered animals to be included in a single breed genetic evaluation.

Growth Traits

The basic BREEDPLAN and GROUP BREEDPLAN model analyses the following growth and growth related traits.

>Birthweight
>200-day weight direct
>200-day weight maternal
>400-day weight
>600-day weight

Being a multitrait model it is not essential to have weights at all ages (birth, 200, 400, and 600-days of age) to calculate Estimated Breeding Values (EBVs) for all of these traits for an individual animal. The 200-day weight maternal trait is an estimate of the milking ability of the dam based on the weaning weight of the calf.

The genetic parameters used in the BREEDPLAN 600 weight analysis are summarised in Table 7.1. The same genetic parameters are used for all breeds. Breed specific parameters are currently being calculated for those breeds which have a data base of adequate size.

Table 7.1 Genetic parameters for traits in BREEDPLAN 600 and in the analysis of male and female reproduction, heritabilities are shown on diagonal, genetic correlations are shown below diagonal. d = direct genetic and m = maternal genetic effect

Weight Category		Birth d	Birth m	200-day d	200-day m	400-day d	600-day d	Scrotal size	Days to Calving
Birth	d	24							
	m	-45	5						
200-day	d	60	0	11					
	m	0	0	0	14				
400-day	d	58	0	60	0	30			
600-day	d	49	0	50	0	70	30		
Scrotal Size		0	0	0	0	24	27	42	
Days to calving		0	0	0	0	-5	0	-29	8

The direct genetic effects of all weights are positively correlated with each other (+0.49 to +0.70). Selection for any weight, thus, also increases genetically the other weights, including birth weight. The maternal effect for 200-day weight is genetically uncorrelated to all other weights. Thus, no EBV can be estimated for the maternal effect for 200-day weight if no 200-day weight has been recorded. The maternal effect for birth weight is negatively correlated to the direct genetic effect for birth weight. This means that animals with high genetic potential for growth in the uterus provide a uterine environment for slower growth for the calves they bear themselves. However, the heritability for maternal birth weight is low (5%).

In 1991 a different model of BREEDPLAN was developed for tropical breeds with traits more applicable to management of northern herds. The 200-day, 550-day and 900-day weights are measurements taken at the end of the wet season while the 700-day weight is measured at the end of the dry season. Separate genetic and environmental parameters are used for Brahman and Zebu crosses, based on estimates obtained by AGBU and CSIRO Rockhampton (Table 7.2).

Table 7.2 Genetic parameters for traits in BREEDPLAN 900, heritabilities on diagonal, genetic correlations below diagonal, d = direct genetic and m = maternal genetic effect. Heritabilities for Zebu crosses are in brackets, for Brahman without brackets

Weight Category		200-day d	200-day m	550-day d	550-day m	700-day d	900-day d
200-day	d	40(25)					
	m	0	5(10)				
550-day	d	80	0	40(30)			
700-day	d	70	0	0		45(40)	
900-day	d	60	0	60		80	45(40)

As with BREEDPLAN 600, there are no genetic correlations of the maternal effect for 200-day weight with the direct genetic effects, and the genetic correlations among the direct genetic effects of all weights are strong and positive. The heritabilities are different for the two genetic types, Brahman and Zebu crosses, with the values for the Zebu crosses being smaller for the direct genetic effects. The heritability for maternal 200-day weight, which mainly reflects the milk production of the cow, is higher for the Zebu crosses, i.e., 10% as compared to 5% for the Brahman.

The weights are allocated to categories on the basis of the mean age of all animals in a contemporary group according to the limits in Table 7.3.

Table 7.3 Age limits in days for allocation of weights to categories in BREEDPLAN 600 and 900

Weight	BREEDPLAN 600	Weight	BREEDPLAN 900
200-day	80 - 300	200-day	80 - 300
400-day	301 - 500	550-day	401 - 700
600-day	501 - 900	700-day	601 - 800
		900-day	701 - 1100

For BREEDPLAN 900 there is an overlap of the limits for 700-day weight with those for 550-day and 900-day weights. To resolve this the breeder has to specify whether the weight was taken at the end of a dry season and, thus is to be treated as a 700-day weight.

Reproduction Traits

Scrotal size and days to calving

The first genetic evaluation for measures of male and female reproductive performance have now been run for some individual British breed herds. Scrotal size is used as a measure of male reproductive performance.

The measurement for female reproductive performance is days to calving, i.e. number of days between the time when the cow is first exposed to a bull, and calving. Figure 7.1 depicts its components.

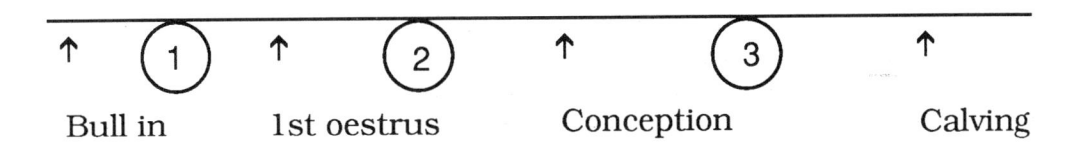

↑　①　↑　②　↑　③　↑
Bull in　　1st oestrus　　Conception　　Calving

Fig. 7.1 The three components which make up days to calving records

The left hand portion of Figure 7.1, the time between the date when the cow is exposed to the bull and the first oestrus after this date, is due to the stage of the oestrus cycle of the cow at the bull-in date. The second portion, from the first oestrus to conception, depends on the ability of the cow to conceive. And the third part is the length of gestation. Given that the first part can be regarded as random noise, and gestation length is rather constant for a given sex of calf, days to calving mainly measures a cow's ability to conceive. Cows that do not conceive are assigned a projected value for days to calving of 380 days. These are included in the analysis, for we expect them to be the poorest genetically for fertility.

The EBVs for scrotal size and days to calving are calculated with a multitrait animal model that also includes the 200-day, 400-day and 600-day weights as correlated traits to account for biases due to selection at earlier weights.

The genetic parameters used in the prediction of breeding values for scrotal size and days to calving are also given in Table 7.1.

The heritability of scrotal size (42%) is in the same magnitude as heritability estimates obtained for weight traits. Only 8% of the variation in days to calving is due to inheritance which is in accordance with heritability estimates found in the literature for other female fertility traits. Higher accuracies on EBVs for days to calving can be obtained when repeated records are observed on a cow or a number of daughters. The repeatability of days to calving records used in the analysis is 0.1. Scrotal size and days to calving are genetically favourably correlated. Thus scrotal size appears to be a useful auxiliary trait to improve female fertility. Scrotal size is also positively correlated (0.24 and 0.27) to 400-day and 600-day weight, while there is a small negative, i.e. favourable genetic correlation between 400-day weight and days to calving.

Gestation length

Gestation length is genetically correlated to birth weight and calving ease, and even if not economically important by itself it is a useful auxiliary trait. It is analysed as a trait of the calf and is moderately heritable (0.22) and also has a small genetic maternal component with a heritability of 0.04 (Table 7.4). It can only be measured in production systems where exact mating dates are recorded; i.e., with AI or hand mating.

EBVs for gestation length are computed for across-herd GROUP BREEDPLAN analyses only, in conjunction with the weights in BREEDPLAN 600. The heritabilities and genetic correlations used in this analysis are in Table 7.5. There is a negative genetic correlation between the direct and the maternal effect of gestation length of -0.41, which is in the same magnitude as the negative genetic correlation between the direct and the maternal effect of birth weight (-0.45). Positive genetic correlations exist between the two direct and maternal effects for both gestation length and birth weight. An animal born after a prolonged gestation due to its genetic effect will also be heavier at birth, but it will bear its offspring for a shorter period and give birth to smaller calves. The negative genetic correlations between the direct genetic effect of gestation length with the post-birth weights means that calves with a genetic potential for long gestation length are expected to be genetically inferior for growth after birth. No genetic relationship exists, however, between the maternal effect of gestation length and any of the post-birth weights.

Table 7.4 Genetic parameters for gestation length and weight traits (heritabilities on diagonal, genetic correlations below diagonal, d = direct and m = maternal genetic effect)

Trait		Gest. Lngth d	Gest. Lngth m	Birth Wt. d	Birth Wt. m	200-day d	200-day m	400-day d	600-day d
Gest. Lngth	d	**22**							
	m	-41	**4**						
Birth Wt.	d	37	-45	**24**					
	m	0	20	-45	**5**				
200-day	d	-11	0	60	0	**11**			
	m	0	0	0	0	0	**14**		
400-day	d	-7	0	58	0	60	0	**30**	
600-day	d	-7	0	49	0	50	0	70	**30**

Carcase Traits

Ultrasonic measurements of rib and rump fat depth and eye muscle area, and the weight of the animal at scanning are used to calculate EBVs for fat depth and eye muscle area. The same genetic and environmental parameters are used for all breeds, but different pre-adjustments for age at scanning using the X-intercept method are applied for Hereford and Poll Hereford, other *bos taurus* breeds and *bos indicus* breeds.

The heritabilities and genetic correlations used for the calculation of carcase EBVs are shown in Table 7.5.

Table 7.5 Heritabilities on the diagonal and genetic correlations below the diagonal for carcase traits

Trait	Scanning Weight	Rump Fat	Rib Fat	Eye Muscle Area
Weight	**47**			
Rump Fat	7	**40**		
Rib Fat	16	84	**35**	
Eye Muscle Area	44	0	5	**21**

The genetic correlations among the traits are positive, with the highest value estimated between the two fat depth measures. Weight and eye muscle area are moderately correlated, indicating that selection for weight also genetically improves eye muscle area.

The Future

Traits that may be evaluated in the future include mature weight and some structural soundness traits such as feet and leg structure and udder and teat soundness. A pelvic size measurement may also be included into a model which provides EBVs for calving ease.

Genetic Evaluation of Calving Ease for Australian Beef Cattle

As presented in Chapter 5 calving ease is scored into 5 categories. For genetic evaluation, the classes 3 and 4 (difficult pull and surgical delivery) are lumped together into one category, and data from category 5 (malpresentation) are excluded. This leaves three categories which are separated on the underlying scale by two thresholds, *viz.*,

1. no assistance
2. easy pull
3. difficult pull and surgical delivery

The statistical model used in the analysis predicts two breeding values for each bull, one for him being the sire of the calf and one for him being the maternal grandsire of the calf. The sire effect predicts half the breeding value for the direct effect, the maternal grandsire effect consists of half the breeding value for the maternal and one quarter of the breeding value for the direct effect (Figure 5.1). The heritabilities for both the direct and the maternal effect were assumed to be 10%, and the genetic correlation between the two effects was 0.5. All relationships among the bulls are taken into account.

In addition to the genetic effects of the sire and the maternal grandsire, the model takes into account the fixed effects of herd-year-season, sex of the calf, and age and grade of the dam.

The breeding values of the sires, as predicted by the model, are expressed in units of the underlying variable. For easier interpretation, they are reported in five categories according to the mean and standard deviation (SD) of the predicted breeding values, *viz.*

A   EBV more than 2 SD above mean
B   EBV 1 to 2 SD above mean
C   EBV 1 SD above to 1 SD below mean
D   EBV 1 to 2 SD below mean
E   EBV less than 2 SD below mean

Chapter 8

Genetic Evaluation in Wool Sheep

Mick Carrick

Introduction

In contrast to the sophisticated national and international across-herd evaluation methods existing in the Dairy and Beef industries, the current methods used in the wool sheep industry are relatively simple and mostly within flock.

Reasons for this relate in part to the low usage of pedigree recording to date in wool sheep and in part to the very large flock sizes. It is not uncommon for a significant stud in the Merino industry to number 10,000 ewes in the breeding flock or even more. The problems of detailed recording and in particular, pedigree recording are therefore about ten times as large as in cattle breeding.

None-the-less, a significant number of Merino sheep breeders are now undertaking such detailed recording and even pedigree recording. The success of this change will also depend in part on the realisation that the creation of elite nucleus flocks in the stud industry, and opening them to female migration from the second tier of the stud, can allow very rapid progress at the same time as reducing the requirement for recording very significantly.

Within Flock Evaluation

Current Woolplan

The most widely available tool for genetic evaluation in the Australian wool sheep breeding industry is Woolplan, an evaluation system designed to operate within the traditional Australian breeding industry in which minimum recording and at most sire pedigrees are kept. In this system, a selection index approach is used, such as outlined in the section on selection objectives. One can choose from a small set of standard indexes or to use one which has been optimised for the particular breeding system in which the flock is operating.

Requirements: At a minimum, the Woolplan system requires that the animal be identified and that at least one measurement of wool weight and fibre diameter at hogget age is recorded. Optionally, the breeder can also record body weight at hogget age and the dam's lifetime number of lambs weaned, and whether the individual was born or reared as a single or twin.

If animals being compared with each other have not been reared together, then management groupings are necessary to prevent identifiable environmental effects from biasing estimated breeding values.

In practice the wool samples taken from the mid-side are sent to an acredited testing house along with the identification of the individual and other data such as greasy wool weight, hogget body weight, birth type, sire if known and age of dam.
Woolplan returns estimates of the breeding values for nominated traits for each animal as well as index rankings for each animal.

The future - Genetic evaluation using BLUP

The acronym stands for Best Linear Unbiased Predictor, and it means that this statistical approach to combining performance information with pedigrees is proven to be the best of the linear ways to use that information. It means also that the prediction of the genetic value of progeny which it produces is unbiased - that any errors are centred around the true breeding value. The mathematics of the system are very complex but the outcome is very simple to use.

The features of this approach to breeding value prediction, sophistication and extreme ease of use to the breeder, have made it the major currency in the beef and dairy industries in which it has been used for a number of years.

When we begin to use a BLUP evaluation system, the first analysis will use the earliest born say 200 lambs as the base against which all subsequent animals are compared. This has the advantage of providing a means of estimating genetic trends from a fixed base allowing across-year analyses to be made. In this analysis the software correctly combines all of the relationships in the pedigree with all of the performance data and any environmental effects which are known. Thus there will be corrections for the age of dams, the birth type (twin or single), paddock or management groups and so on. Then all the equations linking the animals, the traits measured and the environmental effect groups are simultaneously solved in such a way that appropriate corrections are applied to all of them.

The resultant breeding values will be expressed as deviations from the average of the first 200 animals for each trait, and in this respect will be similar in appearance to Woolplan breeding values. However after a number of years in a well designed program, progress will be apparent. The analysis will not only show breeding values for all animals in the flock or linked flocks but also the separate average environmental effect for each year and the average breeding value of the flock for each of the analysed traits. Because both sires and dams are used over a number of years they function as an inbuilt genetic control to allow such trends to be computed in an unbiased way. This not only has tremendous potential as a marketing tool for a breeding system such as ours, it also serves to allow close monitoring of the success of the program at each site.

Not all the estimates have the same accuracy, even though they are the best available at the time of each analysis. The first breeding value estimate one gets for an animal is strongly based on its own performance with information coming also from its parents, grandparents, full and half siblings as well as more peripheral relatives such as uncles, aunts, cousins etc. As time goes on each animal carried on in the flock accumulates more relatives - progeny, grand progeny and so on. The most complete genetic information will exist for the oldest sires perhaps even long dead. In other words the system automatically creates progeny test information for all sires and dams and embryo donors. It does this without the necessity of randomly allocating females to each sire because the simultaneous solution of all the equations correctly adjusts or unbiases the estimates for each animal. We can use special matings and not adversely affect the results.

This feature is important because it allows us to use sensible breeding practices such as mating the best to the best without affecting the accuracy of the breeding value estimates.

The breeding values are easy to use because they are unbiased predictions of each animal's true breeding value and because they behave as though they have a heritability of one. That is if we mate sires whose breeding value is say, plus 1 kg of clean fleece weight with a group of females whose breeding values average plus 0.2 kg, the progeny will have an average breeding value for clean fleece weight of $(1 + 0.2/2) = + 0.6$ kg.

Across Flock Evaluation

The introduction of BLUP methods allows link sires, or dams if these migrate, to provide genetic links permitting a common evaluation across all linked flocks. Similar fixed environmental and some grouped genetic effects can be accounted for in across flock BLUP systems as are curently dealt with in beef cattle and dairy applications.

Two types of across-flock evaluation have been tested in Australia: Central tests and Sire Reference on-farm tests.

The Sire Reference Evaluation was run between 1984 and 1990 by the Western Australian Department of Agriculture who also included a central test for reference sires in the last few years. Unfortunatley funding for this project, which had spread in influence across the Country, has now ceased. The principle behind this test was the use of at least two of a small group of reference sires whose semen was made available at a discount to participating flocks. Statistical analysis of these data allowed correction to be made for between farm environmental effects. The sire reference system allowed many more sires to be evaluated than Central tests and allowed in principle, every progeny test, both central and on-farm to be linked in a single analysis (Swan *et al.* 1992).

Central tests have been run at Hay, Deniliquin, Dubbo and Walcha by the University of NSW, beginning in 1987. These are run on a semi-commercial basis; each sire being included at a fee designed to cover the cost of the test.

Principles in both tests are similar and data may be combined if these principles are carefully adhered to:

- Random allocation of ewes to sire groups.
- Carefully equivalent management of ewes at any time during which they must be separated into groups, mating and lambing, and run together at all other times.
- A minimum number of progeny per sire should be aimed at, usually at least 30.
- An agreed minimum age at which data are collected.
- Agreement on which traits are to be recorded and agreement on standards for each. For example hogget shearing should be the same time after a lamb shearing or alternately there should be agreement on there being no lamb shearing.
- Adequate genetic links established between flocks.

Advantages and Disadvantages of Central and On-farm Tests.

Briefly, Central tests have the advantages that it is easier to ensure that agreed design criteria are adhered to and that progeny are run in the same feed and other environment. Central tests also provide an excellent venue for extension activities related to genetic evaluation.

The major disadvantage of central tests is the high cost of the test and consequently the smaller number of sires tested.

On-farm tests have the advantages that many more sires can be included because of the lower cost, that the only limit to the number of studs which can participate is the semen supply and, by using the best evaluated sires as the next references, rates of genetic change can be improved.

The main disadvantages of on-farm tests are that it is more difficult to ensure that guidelines are adhered to, that timing of data collection may be problematical and it may exclude closed studs.

Reference

Swan AA, Woolaston RR and Piper LR (1992) Establishing a Centralised Database for Merino Sire Evaluation Schemes. Proc 10th Conf AAABG, Rockhampton (In press).

Chapter 9

Genetic Evaluation in Meat Sheep

Robert Banks

Genetic evaluations for the meat sheep industry in Australia are provided through LAMBPLAN. The evaluations are delivered in the form of Estimated Breeding Values and Selection Indices, and are for a range of traits appropriate to the production systems used in Australia and the roles of the breeds used to produce lamb. At this stage LAMBPLAN is aimed entirely at the lamb sector of the sheep meat industry.

The Australian lamb industry makes use of distinct breeds in the terminal sire and maternal roles, and the breeding objectives and hence selection criteria reflect the different roles. The core evaluations are for growth rate; weight at constant age and leanness; fat depth at constant weight, with wool weight and reproduction added for maternal breeds.

This Chapter outlines:

- the traits for which EBVs are reported,
- the selection criteria used to predict EBVs, and the measurement and collection procedures used to obtain them
- the statistical models used for the prediction,
- Selection Indexes for terminal and maternal breeds,
- integration with across-flock genetic evaluations.

Evaluation Structure in LAMBPLAN

The overall evaluation structure within LAMBPLAN is:

Terminal Sires: **Maternal (Crossing and Dual-Purpose):**

Standard: **Standard:**

    Weight   Weight
    Fat Depth   Fat Depth
                            Reproduction
                            Wool Weight

           Optional: Optional:

           Eye Muscle Area   Eye Muscle Area
           Reproduction   Fibre Diameter

The selection criteria for each trait are as follows:
- **Weight:** This is measured as live weight at ages ranging from 150 - 550 days, on both young rams and ewes.
- **Fat Depth:** This is measured as ultrasonic fat depth at the 'C' site, i.e. 45 mm from the backbone at the 12th rib. This measurement must be taken by an Accredited LAMBPLAN Operator.
- **Reproduction:** Three criteria are available and may be used; the animal's own birth and rearing record (single, twin etc), the lambing record of the animal's dam, and the scrotal circumference of the animal if a ram. All animals measured through LAMBPLAN must have birth and rearing records, the dam's record and scrotal circumference are used if available. Scrotal circumference is measured by Accredited LAMBPLAN Operators.
- **Wool Weight:** The animal's hogget (12 months of age) or first adult Greasy Fleece Weight is used, collected by the breeder.
- **Eye Muscle Area:** The prediction is based on the animal's eye muscle depth usually measured at the same time as the fat depth. Real-time ultrasonic equipment is used, and can measure both fat and eye muscle depth, or just the muscle depth.
- **Fibre Diameter:** Measured on mid-side samples of the hogget or first adult shearing fleece.

Statistical Models for Genetic Prediction

Two basic models are used within LAMBPLAN for prediction of EBVs:
- for weight, wool weight, scrotal circumference, and fibre diameter, fitting age of dam, birth and rearing type, and a covariate for age in days as fixed effects, and incorporating relatives' information through multi-trait Animal Model BLUP procedures,
- for eye muscle area and fat depth, fitting weight as a covariate and incorporating relatives' information through multi-trait Animal Model BLUP procedures.

Thus EBVs for Weight, Wool Weight, and Fibre Diameter are at constant age. EBVs for Fat Depth and Eye Muscle Area are at constant weight. The genetic information on scrotal circumference is combined with that obtained from the dam's lambing record and the animal's own birth and rearing status to provide an EBV for reproduction, as lambs weaned per ewe joined.

Selection Indices

A combination of desired-gains and more formal approaches to developing breeding objectives and hence selection indices is used in LAMBPLAN, because of uncertainties in two areas:

- determining the value of reducing fatness in slaughter lambs is made very difficult by the confused marketing signals. Approaches based on estimating the value of kg total body fat have been attempted in other countries but have so far not been applied in Australia.
- the physiological value of body fat reserves to pregnant and lactating ewes is not quantified, but reducing subcutaneous fat could impair maternal performance.

Given these uncertainties LAMBPLAN has used a series of Index options that are most simply described by the relative emphasis placed on different traits, relative emphasis calculated as predicted response in each trait in genetic standard deviation units. These options are:

Breed Group	Relative Emphasis on			
	Weight	Fat Depth	Wool	Reproduction
Terminal Sire:				
High Growth	100	0	-	-
High Lean	0	100	-	-
Lean Growth	50	50	-	-
Maternal:				
Standard	35	30	15	20
Breeders	40	5	30	25

NB: - indicates EBVs can be provided for these traits but they are not included in a Selection Index at this stage.

Eye Muscle Area and Fibre Diameter are not included here, but are also available as options, without being included in Indices.

A number of points should be made about these options:

- the LEAN GROWTH and BREEDERS options have been the most popular to date,
- this very simple approach to both producing Indices and describing reflects in part a very low prior level of performance recording in the Australian meat sheep industry. A simple biological target approach has enormously assisted adoption. At the same time, more detailed analysis as lamb trading in Australia becomes more objective has not revealed any technical deficiencies of this approach: the Index options so far seem to be biologically and economically sensible,
- as lamb trading becomes more objective and better price information becomes available, this indexing system will be modified to incorporate it.

Across-flock Genetic Evaluations

In both the terminal sire and maternal breeds sectors, genetic or pedigree linkage between breeding flocks is significant, despite almost no use of AI. The linkage that exists reflects a high level of migration of sires between breeding flock. This linkage provides a basis for across-flock genetic evaluation through sire referencing. The first across-flock evaluations of this sort for the Australian lamb industry were performed in 1992.

An additional method of across-flock evaluation currently in use in Australia, at least for terminal sires, is through central progeny testing. A number of test stations are being used, and in each case 3-way cross lambs are bred from industry terminal sires of any breed. The primary purpose behind this industry-funded program has been to promote use of better sires in the commercial lamb production sector, but the data sets resulting from this program will be incorporated into across-flock within-breed evaluations.

Finally, LAMBPLAN coordinates breed evaluation for the lamb industry in Australia, through both central progeny testing and use of on-farm records. Currently such evaluations are focussed on the American Suffolk and the Texel. Evaluations of both currently available and exotic maternal breeds, such as the Finnish Landrace, are planned for the period 1995-2000.

Delivering Genetic Evaluations through LAMBPLAN

The requirement that the fat and eye muscle depth measures be collected by accredited field operators combined with the availability of powerful micro-computers has lead LAMBPLAN to be based on decentralised processing of data for within-flock evaluations. These are performed either on farm or at the home base of the operators. This has proven to have considerable educational value, which is particularly important in an industry with almost no prior use of objective recording methods.

Data from within-flock evaluations is collated for processing at a central location for across-flock evaluations, and for research purposes. One planned development within LAMBPLAN is the complete integration of within- and across-flock genetic evaluations.

At 1992, 3 years after the launch of the system, LAMBPLAN is testing 50% of the annual terminal sire intake of the lamb industry and some 5% of the intake of maternal rams. Acceleration of improvements in lamb trading and objectivity within the processing sector appear to be encouraging the growth of performance recording and genetic evaluation through LAMBPLAN.

Summary

LAMBPLAN is designed to be a complete genetic evaluation and improvement system for the Australian lamb industry. The overall structure and mode of delivery are simple, reflecting the low prior usage of objective methods. At the same time, the availability of pedigree information allows use of BLUP technology without an additional recording burden on stud breeders.

The system aims to provide information at two levels for two classes of traits:

- evaluation of different breeds
- evaluation within breeds, both within- and across-flock
- evaluation of growth phase traits: growth rate, leanness and muscling
- evaluation of maternal phase traits: reproductive performance and wool cut.

The fact that market signals for several traits are unclear has encouraged adoption of a desired-gains approach to indexing. This approach has been well-accepted by breeders and is supported by the existence of clearly defined and recognised terminal sire and maternal breeds, with different objectives in biological terms.

Chapter 10

Genetic Evaluation in the Dairy Industry

Mike Goddard

Many countries now use an Animal Model BLUP analysis to estimate breeding values for dairy cattle. More information on this statistical method can be found in Schmidt (1988). The Australian analysis is described briefly in Jones (1985) and Jones and Goddard (1990).

Calculation of EBVs for Milk Production Traits

The performance or phenotype of a cow depends on her genetic merit or breeding value and environmental factors. In estimating breeding value for milk yield we attempt to correct for these environmental factors in two ways. Firstly, we adjust for known environmental effects. In most countries a 305 day lactation yield is calculated for each cow and adjusted for factors such as cow age. In Australia individual test-day yields are adjusted for the age of the cow and her stage of lactation. Her adjusted test-day yields are expressed as a deviation from herd test-day mean and combined to give a total lactation yield. This lactation yield is similar to the production index (PI) calculated by many herd recording bodies but is expressed in kilogram of fat and protein and litres of milk.

Secondly, lactation yields are only compared among cows calving in the same herd in the same year and season. This is done by using the statistical model

$$y = hys + a + p + e$$

where

y = adjusted lactation yields
hys = effect of the herd-year-season to which the record belongs
a = breeding value of the cow
p = a permanent environment effect which affects all lactations of the cow
e = a temporary environment effect which affects only this lactation.

BLUP is used to simultaneously estimate the **hys**, **a** and **p** effects. The estimate of **a** becomes the EBV for that animal. Some countries publish Predicted Transmitting Abilities (PTAs) which are half the EBV.

In calculating the EBVs notice is taken of the relationships between animals. Consequently the EBV of a cow reflects her own milk yield and that of her relatives. Naturally the EBV of a bull is based entirely on the performance of his relatives. The most important relatives are his daughters because they measure his breeding value directly.

Calculating the estimated breeding value of all animals, male and female, in this way is referred to as an 'Animal Model'. Australia, in 1984, was the first country to use an animal model BLUP for national dairy evaluations. Many countries now use an animal model although a sire model is still used in some countries.

Other features of the analysis of production traits are described below. These features apply specifically to the Australian analysis but other countries use similar methods.

Comparison Across Breeds, Herds and Age-Groups

The difference in yield between heifers and mature cows may vary from herd to herd. Consequently common age corrections might only apply to an average herd. Therefore, different age groups within the herd are treated as separate herd-year-seasons. This means that the milk yields of cows of very different ages are not compared.

Breeds are analysed separately except for the red breeds which are analysed together. A cow with a Holstein sire is in the Holstein analyses, a cow with a Jersey sire in the Jersey analysis. Within a breed, purebred and crossbred cows are placed in separate herd-year-seasons, so that their milk yields are not directly compared.

The widespread use of bulls by AI creates links between herds so that the EBVs can be compared across herds within a breed. However, EBVs cannot be compared across breeds.

Weighting and Standardisation of Yields

Each trait is analysed separately, i.e. a single trait BLUP.

Lactations in progress are used provided there are at least two test days. A major advantage of using test-day information is that lactations in progress are used efficiently.

Lactation yields are weighted according to how accurately they reflect breeding value. First lactations have the highest weight and later lactations get progressively less weight. Lactations with a small number of test days get less weight than complete lactations.

In some herds the variation in milk yields is much higher than in other herds. This could cause the best cow in a variable herd to get a higher EBV than an equally good cow in a lowly variable herd. To avoid this, all herds are standardised to have the same variability.

The Base

Breeding values can only be described by comparing one animal with another. In calculating EBVs all animals are compared to a group of bulls called 'the base'. The base is the average breeding value of a group of widely used AI bulls which had daughters on file in 1981/82. The base is given an EBV of zero and the EBVs of all other bulls and cows are expressed relative to this base. Change to a new base would merely move all EBVs up or down by a constant amount without changing the ranking of animals.

Policies for setting a base vary widely around the world. In the USA the base is currently the average of cows born in 1985 but the intention is to change the base every 5 years. In Canada the base is rolled over every year so that the average remains zero.

Overseas Bulls

Different countries use different scales for expressing estimated breeding values so that bulls with evaluations in different countries cannot be directly compared. Goddard (1985) showed how sire evaluations in one country can be converted to the scale used by another country. In Australia bulls from USA, Canada and NZ have their home country EBV converted to the Australian Breeding Value scale so that all bulls can be readily compared by Australian dairy farmers. When the bull produces daughters in Australia, this converted breeding value is incorporated into the analysis so that the bull obtains an EBV which reflects the performance of his Australian and overseas daughters. The reliability of a converted breeding value is less than that of the bulls proof in his country of origin to allow for the uncertainty in converting between countries. For the same reason, the effective number of overseas daughters in an Australian EBV is limited to about 30. Consequently when a bull has a large number of Australian daughters these dominate the EBV.

Reliability

The EBV of a bull or cow is the best estimate of its breeding value given the information available. Unless a bull has a large number of daughters, his EBV will not estimate his true breeding value exactly. However, an important property of BLUP estimates is that for 50% of bulls and cows the true breeding value will be less than the EBV and for 50% of animals the true breeding value will be greater than the EBV. Consequently the average EBV of a group of animals will be close to their average true breeding value.

The amount by which the EBV is likely to differ from the true breeding value (i.e. the likely error) is measured by the 'reliability'. The reliability is the square of the correlation between EBV and true breeding value, i.e. the square of the accuracy as normally calculated. For a group of bulls, all with reliability 81%, the correlation between their EBV and true breeding values in 0.9 ($0.9^2 = 0.81$). Therefore selection on these EBVs is, in general, quite accurate. However, for individual bulls the error in their EBV can be appreciable. For instance, for the fat EBVs of this group of bulls, the average error is 0 (some positive, some negative) but the standard deviation of errors is 6kg. Consequently 1 bull in 20 will have an EBV for fat which is 12kg in error. As this bull gets more daughters his EBV will move towards his true breeding value.

The reliability of a bull's EBV depends mainly on the number of daughters he has. The reliability for fat, protein and milk is approximately $n/(n+15)$ where n is the number of effective daughters the bull has. Thus a bull with 60 effective daughters has a reliability of 80%. A bull's effective number of daughters is usually only about 2/3 his actual number of daughters because some daughters are in herd-year-seasons with only a small number of contemporaries against which to compare them. Information from the EBV of a bull's sire and dam also contribute to his EBV and increase the reliability slightly.

Cows seldom have a large number of daughters so the reliability of their EBVs is usually low compared to that of proven bulls. It increases with the number of lactations the cow has, and with the reliability of the EBV of her sire and dam but is usually below 50%.

The errors in EBVs discussed above are the inevitable statistical errors that occur because all milk yields are affected by random environmental factors and because we only have available a sample of all a bull's possible daughters. If the data from which EBVs are calculated is not sound other errors will be generated. For instance, if a cow is given preferential treatment this will bias her EBV. If several daughters of a bull receive preferential treatment it will also bias the bull's EBV. This should not happen in a well designed progeny test program because a bull's daughters will be spread over many herds which have no financial interest in the sire.

Errors in pedigree information can also generate errors in EBVs. If a cow's sire is wrongly recorded it causes a bias to her EBV and a small bias to the EBV of the incorrect sire.

Tests for bias in EBVs suggest that these sources of error are not common but we rely on all those entering data into the system to continue to exercise care concerning the quality of the data.

Interpretation of EBVs

The EBV of a calf is the average EBV of its sire and dam. If we have a cow with protein EBV = 10 kg and we mate her to a Bull A with protein EBV = 30kg, the EBV of the newborn calf is $\frac{1}{2}(30+10) = 20$kg. If we had used Bull B (EBV = 10kg) the calf's EBV would be $\frac{1}{2}(10 + 10) = 10$kg. The expected difference in protein yield between these calves is the same as the difference in EBV i.e. 20-10 = 10kg. Thus bulls which differ by 20kg in EBV produce daughters which differ by 10kg, because the daughters get half their genes from their sire and half from their dams.

EBVs for young bulls with no daughters can be calculated in the same way i.e. $\frac{1}{2}$ (sire's EBV + dam's EBV). A desirable feature of BLUP estimates of breeding value is that they allow all animals to be compared. You can directly compare the EBV of young progeny test bulls with that of proven bulls and select the one offering the best value for money.

Although the EBV of a sire predicts the performance of the average daughter very well, individual daughters vary widely. All bulls produce some bad daughters. However, if a dairy farmer breeds his herd from high EBV bulls, his herd average will be higher than if he used low EBV bulls.

Use of EBVs

Bull EBVs are used by dairy farmers when selecting semen to buy and by AI studs when selecting the sires of young bulls. A method by which the dairy farmer can get the best value for money, considering all traits, is considered in the chapter on selection objectives.
If a dairy farmer uses a single bull very widely he runs a risk that the bull's true breeding value is substantially less than his EBV. To minimise this risk a dairy farmer should always use a group of bulls (e.g. 4 or 5).

The lower the reliability of a bull's EBV the larger are the possible errors, up or down, is his EBV. Therefore unproven bulls should always be used as a team. The average EBV of a team of young bulls will be close to the average true breeding value.

Bulls which have been progeny tested in a small number of herds represent an additional risk, due to the possibility that their daughters may have received preferential treatment. In Australia the number of herds in which a bull has daughters and the number of daughters in the 2 herds with the most daughters are published. This enables farmers to identify and, if they wish, avoid bulls that don't have daughters spread across many herds.

Cow EBVs can be used by dairy farmers to select cows from which they will keep a replacement heifer. Unfortunately, in seasonal calving herds, most dairy farmers have little opportunity to carry out such selection because they need all AI bred heifer calves.

If a dairy farmer is selecting which of his yearling heifers to keep, or which heifers of other farmers to buy, he could use their EBV = $\frac{1}{2}$(EBV sire + EBV dam) to help make decisions.

Cow EBVs are not intended as a culling guide. A cow may have a low EBV, although she is an average milker, because she has a poor pedigree. In deciding whether or not to retain the cow it is her own production which is most important. But, in deciding whether or not to keep a heifer calf from that cow, it is her breeding value that is relevant. Her EBV takes into account her own production and her pedigree to estimate her breeding value.

Traits Other Than Production

The other traits for which EBVs are calculated in Australia are length of herd life or survival, temperament, milking speed, likeability, calving ease and 30 type or conformation traits. These are all done as single trait analyses using a sire and maternal grand sire model instead of an animal model.

Survival is the ability of cows to survive in the herd from one year to the next i.e. not to be culled or die. The Australian average is about 83% of cows survive or 17% are culled or die each year. EBVs for survival are expressed as deviations from the average. A bull with an EBV of +6% will have 3% more daughters surviving each year than a bull with EBV = 0. The cow gets half her genes from her sire. High survival EBV means a high average length of life in the milking herd for his daughters.

Survival is a trait of low heritability because it is influenced by non-genetic factors. Consequently a bull needs many daughters in order to have a highly reliable EBV. Also as his daughters age and have more opportunities to either survive or not, the reliability of his survival EBV increases.

The workability traits, temperament, milking speed and likeability, are scored by dairy farmers on a 5 point scale from very good to very bad. The two lowest grades represent cows which the farmer thinks unsatisfactory for the trait. About 12% of cows are scored as unsatisfactory for each trait. Sire evaluations for these traits are expressed as the percentage of a bull's daughters which are satisfactory. These are not estimated breeding values but predicted transmitting abilities. If Bull A has a Predicted Transmitting Ability (PTA) for temperament of 90% then he will have 4% more satisfactory daughters than Bull B with PTA = 86%. You don't have to half the difference because they are already expressed as transmitting abilities.

The data for calving ease analysis comes from farmers who record on the herd recording forms whether a birth was unassisted or difficult. The PTAs for calving ease give the predicted percentage of difficult calvings in mature cows. Unfortunately there is not enough data to base the PTAs on heifers where difficult calvings are a greater problem. For instance, a bull with PTA = 6% will cause 2% more difficult calvings than a bull with PTA = 4%. The average for Holsteins is approximately 4%.

In heifers the proportion of assisted births is 30% and, although low calving difficulty Holstein sires reduce the proportion needing assistance, they do not produce trouble free heifer calvings.

Type data comes from classification of registered and progeny test herds by breed society classifiers, mainly from the Holstein-Friesian Association of Australia. The EBVs are scaled so that their standard deviation for each trait is 5 and mean 0. Therefore bulls vary from approximately -10 to +10. This EBV scale is quite distinct from the scale on which the cows are scored by the assessors.

Australian Dairy Herd Improvement Scheme (ADHIS)

Genetic evaluation for the Australian dairy industry is carried out by ADHIS which is the responsibility of the Australian Dairy Farmers Federation and is funded by the Dairy Research and Development Corporation. ADHIS is administered by an executive officer and the scientific work is carried out by the Victorian Department of Food and Agriculture. Their major task is to calculate EBVs or PTAs, which are called Australian Breeding Values or ABVs, for Australian dairy cows and bulls. To do this they rely on data supplied to ADHIS by the state herd recording organisations, by AI studs and by breed associations.

Bull ABVs which meet reliability standards are published by ADHIS. A list of all bull ABVs is available as computer readable files or as computer printouts. Cow ABVs are returned to state herd recording organisations who notify dairy farmers of the EBVs of their own cows.

Extension Problems

Below are some questions that are frequently asked about EBVs.

Why didn't my cow Flossy get an EBV?

Cows which have an unknown sire are not included in the analysis. This could be because no sire was entered on the herd recording data, because the sire does not exist on the state bull file or because, according to ADHIS records, semen was not available from that sire at the time the cow would have been conceived. Cows are also excluded if they are missing vital information, such as date of birth, or if they have no valid production data.

Cow 1379 produced 200kg of fat last year while 1721 produced only 150kg, yet 1721 got the higher EBV. How can this be?

There are a number of reasons why a cow's EBV differs from her raw production.

Firstly, when calculating an EBV production figures are corrected for age and stage of lactation. The effect of corrections can be checked by looking at the cows' production index if they are available.

Secondly, the production of each cow is compared to the mean of their herd-year-season. Cow 1379 may have been one of a small number of Autumn calvers in the herd who had a higher average production than the Spring calvers.

Thirdly, a cow's relatives are considered in calculating her EBV. A cow may perform well herself despite genetically inferior parents because she received, by chance, favourable environmental circumstances. However, these favourable environmental effects will not be passed on to her offspring, but the poor genes of her parents will be. It is the genetic merit she will pass on to her offspring that the EBV is designed to predict.

The relative weight given to her own production and her pedigree in calculating her EBV is controlled by the heritability of the trait. For a trait with a high heritability an animal's own performance is a good guide to her breeding value. The heritability ADHIS uses for milk, fat and protein yields is 25%. This is a fairly standard figure used world wide and based on many analyses of field data.

Why did the EBV of bull ABCD drop 10kg between 1990 and 1991?

EBVs change mainly because more information becomes available e.g. more daughters of the bull are included or existing daughters have more lactations. As more information accrues a bull's EBV will tend to move towards his true breeding value. Some bulls will move up and some down. For most bulls with a reliability over 70% these changes are not great. However, as discussed earlier, a small number of bulls will change by as much as 12kg for fat EBV.

Because bulls are equally likely to increase or decrease in EBV, the average of a group of bulls is unlikely to change very much. In other words, on average what you see in an EBV is what you get.

References

Goddard ME (1985) A method of comparing sires evaluated in different countries. Live Prod Sci 13: 321-331

Jones LP (1985) Australian Breeding Values for Production characters. Proc 5th Conf AAABG 242-247

Jones LP and Goddard ME (1990) Five years experience with the animal model for dairy evaluations in Australia. Proc 4th Wrld Congr Genet Appl Live Prod 13: 382-385

Schmidt GH (1988) Proceedings of the animal model workshop J Dairy Sci 71: suppl.2

Chapter 11

Genetic Evaluation in the Pig Industry

Tom Long

Introduction

Genetic evaluation can be thought of as an assessment of breeding programs to evaluate genetic progress being made, or evaluation of individual animals for their value as a parent (breeding value). The type of genetic evaluation pig producers do, depends on which sector of the industry they operate in. Figure 11.1 is a simplified diagram of the structure of the pig industry with regard to genetic improvement.

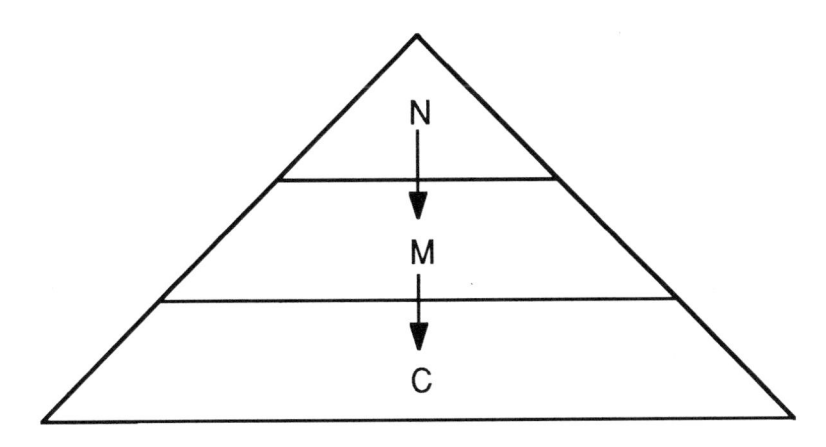

Fig. 11.1   Pig industry structure with regard to genetic improvement
N = Nucleus; M = Multiplier; C = Commercial.

At the top of the pyramid, the nucleus sector is where genetic progress must be achieved and passed downward to the rest of the Industry. At this level evaluation of individual animals for genetic merit is the key to making selection and culling decisions to elevate the average breeding value of the herd. Also in this level, evaluation of the breeding program is important as this assessment tells breeders how effective their selection procedures have been in moving toward their breeding goals.

Producers in the multiplier level are coupled with the genetic improvement program of the nucleus breeders, either through a breeding company framework or through buying boars from elite breeders. Their genetic evaluation is mainly: evaluating breeding programs of different nucleus breeders to choose who to get seedstock from or choose which breeding company to work with, and maintaining quality control, i.e. multiplying and passing on the genetic improvement attained in the nucleus to the commercial sector.

Commercial producers are breeding pigs only for market and, since crossbreeding is used extensively, are buying their additive genetic improvement from other seedstock producers. They may be raising their own gilts as replacement females, but the majority of the genetic improvement they are getting is from the males that they introduce into the herd. Therefore, genetic evaluation for these commercial producers mainly entails evaluating the genetic, and health programs of various seedstock producers, and choosing the one whose program is making good genetic progress with animals or semen available of superior genetic merit.

Although the major focus of this chapter will be genetic evaluation of individual animals for breeding value in the nucleus sector, it must be emphasised that genetic evaluation of breeding programs occurs in all three sectors of the Pig Industry, and is important in the genetic improvement the Industry attains as a whole.

Traits of Economic Importance

Traits of economic importance to the Pig Industry and their heritabilities are presented in Table 11.1. The heritabilities given in this table are average values, and these can vary depending on the population being considered. As a general rule, however, reproductive traits tend to have low heritabilities, production traits such as growth rate or feed efficiency tend to have moderate heritabilities and carcase traits tend to have high heritabilities.

Table 11.1 Economically Important Traits and their Heritabilities

Trait	Heritability, %
Number of pigs farrowed per litter	10
Number of pigs weaned per litter	10
Birth weight	20
Weaning weight	20
Feed efficiency	25
Growth rate	30
Age at puberty	35
Backfat	40

Whether these traits are included as selection criteria in a breeding program will depend on the breeding objectives of each individual breeder, i.e. whether animals being evaluated are from a maternal line, a terminal sire line or a dual purpose line of pigs. Other traits that are currently being considered as selection criteria for inclusion in breeding objectives are: lean meat yield, meat quality, appetite and mothering ability of sows.

On-Farm Testing

For efficient genetic evaluation of animals to take place, objective measures of selection criteria are required for animals which are being considered for selection. These measures can be taken either on-farm or in a central testing station. One of the main advantages of on-farm performance testing is that many more animals can be tested than in central testing schemes. This can increase selection differentials and also is useful in assessing the genetic and environmental trends of the herd to evaluate genetic progress being made in the herd. Traits usually measured are average daily gain and backfat on animals in the grower herd and number of pigs born alive for sows. Also, some breeders are obtaining feed efficiency data on-farm, although this can be costly, and most breeders rely on central test stations to obtain this type of measurement. Traits measured will vary from breeder to breeder depending on how they define their breeding objectives and resources available to do on-farm testing.

On-farm testing programs should seek to test as many animals as possible. Recording only data from animals that have been selected is not performance testing. Also, testing groups should be as large as possible. Different methods of genetic evaluation try to account for environmental effects differently, but testing groups should be relatively large to accommodate this. Large test groups occur normally in large breeding herds, and smaller breeders should try to assure large test group size by farrowing sows in groups.

One of the main disadvantages to on-farm testing is that, due to the large variability in environments between farms, it is difficult to compare performance figures or breeding values of animals in different herds. Central testing is a scheme that tries to solve this dilemma.

Central Testing

The idea behind central testing is to bring animals in from a range of environments (farms) and test them in relatively uniform environmental conditions so genetic comparisons between these animals can be made. If large numbers of animals could be tested, genetic differences between farms could be assessed. Unfortunately, central test stations have size limitations, so can only serve as a supplement to good on-farm performance testing programs. Boars which do well in a central test can be put into an AI centre where semen is available to a number of producers. This gives other breeders access to animals that have done well in the central test without the health risk involved in introducing live animals to a herd. For commercial producers this can reduce the genetic lag in passing improvement to this sector as the multiplier sector is circumvented and improvement goes from the nucleus to the commercial sector. Some of the problems with central testing include: ineffective sampling of boars to become candidates in the central test, health risks involved in pooling boars from a number of farms, limitations to the number of animals that can be tested and high performing boars or their semen not being used back in their herd of origin. Many breeders use the central test only as a marketing tool rather than as a tool for genetic improvement, so genetic links between boars put into a central test and animals on-farm could be weak.

Methods of Evaluation

Once measurements or data are taken on a group of pigs, these animals need to be evaluated for their genetic worth as a potential parent (breeding value) so selection and culling decisions can be made. There are a number of methods of genetic evaluation being used in the Pig Industry, and this section will describe those methods.

Visual appraisal

This method of genetic evaluation entails viewing the animals and making selection decisions on how they look. Arguably, some visual appraisal of potential breeding stock is necessary to evaluate breeding soundness, e.g. feet and legs, teat number, genitalia. One pig breeder has defined a sound pig as one who 'has 12 good teats and can walk'. Also, some breeders have customers who demand a specific type (conformation) of pig, so feel they need to apply some selection pressure to type. Unfortunately, there are some breeders whose only genetic evaluation of their pigs is visual appraisal, and this is not a very effective method to maximise genetic progress in a herd. Many of the visual traits are not highly correlated to production traits of economic significance. Visual appraisal must be coupled with objective measurement of animals and evaluation of those measurements in developing a good breeding program.

Single trait selection

With this method the breeder has decided there is only one trait of economic importance, e.g. growth rate in defining a breeding objective and is putting all selection emphasis on that trait. Evaluation is simple, requiring one measure per animal, ranking them within the test group and selecting the best. A breeder can make good genetic progress using single trait selection, depending on the heritability of the trait, and there are some niche markets for breeding stock that could be exploited using this method. Single trait selection, however, ignores the fact that there is more than one trait of economic importance to the Pig Industry and does not put any selection pressure on these other traits, except when they are genetically correlated to the trait being selected for. If these genetic correlations are favourable then some genetic progress can be made in those other traits, but this progress would be sub-optimal to other evaluation methods. Also, if antagonistic genetic correlations exist with other traits of economic importance, reduced performance in these traits could offset the improvement made in the trait being selected for. Using single trait selection can make breeders susceptible to falling into the trap of selecting for one trait for a few years, seeing performance fall in other traits, changing selection criteria to other traits and never moving effectively towards an optimal breeding goal. Methods which combine more than one trait are more effective in developing a good breeding program.

Selection Index

Selection indexes use heritabilities, genetic correlations and economic weights to combine several traits into a single index value upon which to base selection. This method has the advantage over single trait selection of being able to put selection pressure on a number of traits. Calculations are relatively easy and can be performed on-farm with a programmable pocket calculator. An example of a selection index being used in the Pig Industry is given below:

$$INDEX = 30 \cdot ADG - P2\ fat + LWT \cdot 0.1$$

where, ADG = average daily gain (grams/day)
P2 fat = fat measurement located 65 mm laterally from the mid line of the back and immediately behind the last rib.
LWT = liveweight at testing

Many different selection indexes are currently being used in the Pig Industry. These indexes differ because either different traits are included, different economic weights for the traits are used and/or different heritabilities and genetic correlations between the traits are used in formulating the indexes. Some breeders are using an index that was derived for the MLC (Meat & Livestock Commission) in the UK. Using this index assumes that genetic parameters for British lines of pigs and the average economic weightings for traits in the British production/marketing system at the time the index was derived are the same as the Australian situation now. Some Australian State Departments of Agriculture have provided on-farm selection indexes in their pig herd improvement schemes for breeders to use. These assume that breeders in the respective state have the same basic breeding objective and economic weightings. Since breeders do have different breeding objectives and production and marketing systems, it would be better if they could develop their own index for their breeding goal, but many do not have this level of expertise. Using inappropriate indexes can give sub-optimal response in moving toward a breeding goal, depending on how much the situation in which the index is being used deviates from the assumptions used in the development of that index.

One thing selection indexes have in common is that comparisons of animals for selection must be done within the test group, e.g. an intake of boars into a central test station or animals coming off test at the same time in an on-farm performance testing program. Selection index does not account for differences between test groups so index values of animals from different test groups cannot be compared without bias. This means that index values of an old boar in the herd and a young boar coming off test cannot be compared for selection/culling decisions, and genetic trends over time cannot be assessed to evaluate the effectiveness of the breeding program. These last two problems are also disadvantages of using single trait selection, as ranking of animals needs to be done within the test group.

BLUP and PIGBLUP

Best Linear Unbiased Prediction (BLUP) is being used routinely in a number of livestock industries and is currently being implemented by the pig industries in a number of countries. The reason BLUP is becoming the method of choice for genetic evaluation is because it has several advantages over other methods of genetic evaluation. These include the following:

- BLUP is a procedure which uses information from all known relatives of an individual, thereby, providing a more accurate prediction of the genetic merit of that animal.
- It facilitates comparisons of genetic merit between animals producing records in different management regimes or over different periods of time.
- It can be used to make comparisons between animals from different herds, providing there are adequate genetic links between those herds.
- It facilitates comparisons of the genetic merit of animals with differing amounts of information, such as a sow with three litters *vs* a gilt that has no farrowing records.
- It allows comparisons to be made among animals that have undergone different amounts of prior selection, such as evaluating males and females for reproductive traits.
- It partitions genetic and non-genetic effects on performance into their respective components, thus enabling breeders to assess genetic change over time.

These features give BLUP a distinct advantage over previous methods of genetic evaluation for pigs, e.g. single trait selection or selection index, particularly by having higher accuracy of estimation and by facilitating genetic comparisons of animals over time, herds and management regimes within herds.

As might be expected, the added power of BLUP has brought with it the potential to create genetic problems if it is not properly used. Since BLUP uses information on all known relatives, the potential exists for an increase in the rate of inbreeding over single trait selection or selection index in the closed herd situation, especially when selection is on a trait of low heritability and records are limited to one sex, such as litter size. Given that relatively high rates of inbreeding can depress phenotypic performance in pigs, inbreeding has the potential to negate some of the advantage of the increased rate of genetic gain from BLUP.

Several authors have proposed ways of addressing this dilemma and some of these are listed below:

- Maintain large numbers of families within the line, especially for maternal lines.
- Include production traits along with reproduction traits in the breeding goal.
- Account for an animal's average relationship to the herd as well as its estimated breeding value (EBV) in making selection decisions.
- Incorporating EBVs and inbreeding of expected progeny into mating decisions and strategies.

The above suggestions give ways of dealing with the accumulation of inbreeding, but it must be stressed that inbreeding in any closed population especially a maternal line, can be a problem, regardless of the method of genetic evaluation. Also, from a practical viewpoint, it should be stated that there are very few totally closed herds in Australia as most breeders occasionally introduce outside boars into their breeding programs.

Another potential disadvantage of using BLUP is that it is computationally more demanding than other selection methods. Single trait selection usually requires only one objective measurement to rank animals, and selection indexes can be calculated on a programmable pocket calculator. BLUP, however, solves a large number of equations simultaneously, and this requires a computer. In the past these types of calculations have had to be done on a mainframe computer, but a computer software system, PIGBLUP, developed by a collaborative team at AGBU, Australia and University of Goettingen in Germany has been developed so BLUP evaluations on pigs can be done with a microcomputer on-farm. The current version of PIGBLUP analyses average daily gain, backfat, feed conversion ratio, and number of pigs born alive. In addition to producing BLUP EBVs for each animal PIGBLUP also produces genetic and environmental trends over time for each of the traits considered. As stated earlier, genetic evaluation is a process of both evaluating individual animals for genetic merit and evaluating the breeding program as a whole and having these trends is important in facilitating that. PIGBLUP also contains an integrated module, $INDEX, which combines the BLUP EBVs into a single $EBV, using a bio-economic profit function, so breeders can establish their own economic objectives and deliver them directly into an index customised for each of their herds.

The method of genetic evaluation producers choose will depend on what sector of the industry they are in, their breeding objectives, resources they have available for testing animals and resources they have available for evaluation procedures. There is a trend, both in Australia and internationally, for there to be fewer and larger pig production units, and, although being larger is not necessarily more efficient, if this trend continues, there will be fewer pig producers operating in a much more competitive marketplace. This higher level of competition, in an already competitive industry, would suggest that producers must keep in mind not only their present situation but where the industry is moving in choosing genetic evaluation procedures if they wish to remain competitive in the future market environment.

Fully Integrated Genetic Evaluation System

In theory, the most powerful genetic information for the Industry would be produced from genetic evaluations across all seedstock producing herds where all herds were strongly linked genetically, via some common sire usage and all feed intake, growth, reproduction, carcase quality and quantity, and survival measures of economic importance were recorded on all potential replacement stock. Unfortunately, this ideal is unachievable because:

- Not all herds will desire to participate in an across-herd analysis for a variety of reasons, e.g. financial, confidentiality.
- Not all herds will be adequately linked genetically.
- Not all animals will have records for all traits.
- Some animals will be recorded in different environments at different stages of their life.

- Some herds will not have data computerised so information from all herds cannot be analysed.

Nevertheless, in establishing guidelines for particular breeding operations, such as central testing, sire referencing, and on-farm testing we should not lose sight of this ideal.

Guide for Consultants

Following is a brief summary of the points to consider while consulting with a producer regarding genetic evaluation.

- Determine what sector of the industry the producer is operating in, nucleus, multiplier or commercial. Many producers selling breeding stock are multipliers. They buy the majority of their replacement boars from other breeders, but think they are an elite or nucleus breeder, so this assessment should be done diplomatically.
- If the producer is commercial, the main genetic evaluation is of the breeding and health programs of potential suppliers of breeding stock.
- If the producer is a multiplier, the questions to consider are where to get replacement stock and how much testing is to be done by the multiplier. Since the multiplier is serving a quality control function with regard to multiplying the genetic improvement attained in the nucleus sector, some testing is required.
- If the producer is a nucleus breeder, the majority of replacement boars are raised on-farm. More detailed questions are required of the consultant to assess the progress already made by the breeder and to suggest possible enhancements to the breeding program.
    - What is their breeding objective, or objectives if they are producing several lines of pigs? If they are unclear in this area, the consultant should work with them first in defining clear breeding objectives prior to considering genetic evaluation. This is analogous to asking first where someone wants to go before asking how they are going to get there.
    - What traits are currently being measured?
    - What recording system is being used?
    - How are they using records in evaluating animals for selection? (e.g. ranking animals for single trait selection, a selection index, or PIGBLUP).
    - Is their system computerised?
    - What is the size of the operation and resources available for testing and evaluation?
    - What is the size of their average test group?
    - Are there different management regimes such as an old *versus* new grower shed or different feeds to different groups of pigs, that should be recorded so they can be accounted for in a genetic evaluation?
    - What are current selection and culling procedures for old and new animals?
- These questions should give the consultant a good picture of progress already made by the breeder so that recommendations can be made to improve their operations.

Chapter 12

Across-breed Genetic Evaluation

Andrew Swan

Across- *versus* Within-breed Evaluation

Chapters 5 to 11 have dealt with prediction of breeding values within breeds using BLUP. It may also be desirable to use BLUP for across-breed genetic evaluation. This allows direct comparison of animals of different straightbred and crossbred types which may be candidates for selection in crossbreeding programs. Current genetic evaluation systems do not allow these types of comparisons. For example, it is not valid to compare an Angus bull with an Hereford bull, even though both may appear in the respective sire summaries for these breeds. In this Chapter, across-breed genetic evaluation methods and their potential application to different livestock industries will be discussed.

Understanding Crossbreeding Effects

Current within-breed models used in genetic evaluation systems cannot be used to analyse across-breed data because they do not account for crossbreeding effects, including breed effects and heterosis. Breed effects are differences in genetic mean between straightbred and crossbred types, and are often referred to as additive or average breed effects, or breed differences. Heterosis, also referred to as hybrid vigour, is observed when crossbred populations are superior to the average of their straightbred parent breeds. For some traits, crossbreeding effects can be further partitioned into direct and maternal components. For example, direct heterosis is the benefit an animal receives because it is crossbred, while maternal heterosis is the benefit received from a crossbred mother.

Crossbreeding effects are often estimated from experimental data using a regression or weighting approach. Firstly, a breed effect is defined for each straightbred type. The breed effect of a crossbred type is then derived by weighting these straightbreed effects by the expected proportion of genes from each breed. For example, consider the F_1 cross between two hypothetical breeds, A and B. Denoting the breed effects as g_A and g_B for breeds A and B respectively, the breed effect of the F_1 is:

$$\tfrac{1}{2} g_A + \tfrac{1}{2} g_B$$

Each effect is weighted by one half, because half the genes in the F_1 come from each breed. If F_1 dams are then backcrossed to breed A sires, the breed effect of the resulting progeny is: $\tfrac{3}{4} g_A + \tfrac{1}{4} g_B$, because on average three quarters of the genes are derived from breed A, and one

111

quarter from breed B. Weightings for breed effects in a variety of straightbred and crossbred types are shown in Table 12.1.

Breeds also differ in average maternal performance for maternally influenced traits, so maternal breed effects can also be defined: m_A and m_B for breeds A and B. The maternal effect for a particular crossbred type is determined by examining the proportions of genes from each breed in the dams. For example, in an F_1 cross with a breed A sire and a breed B dam, the maternal effect for a trait is m_B, because the dam is a breed B straightbred. In the backcross to breed A described above, the maternal effect is $\frac{1}{2}m_A + \frac{1}{2}m_B$, because the dam is an F_1 with half her genes for maternal performance derived from each breed.

Heterosis is often explained using the dominance theory. Under the dominance theory, the loss of heterosis is expected to be linear with the loss of breed heterozygosity. That is, if breed heterozygosity is reduced by half, heterosis is expected to be reduced by half. Breed heterozygosity occurs at a locus when the genes from the sire and dam originate from different breeds. These concepts are explained more fully in Figure 12.1; for each straightbred or crossbred type, an imaginary chromosome with four loci is depicted. In a breed A individual, the genes contributed by both sire and dam originate from breed A. Therefore, there is no breed heterozygosity and no heterosis expressed. In an F_1 cross, all genes from the sire originate from breed A, and all genes from the dam originate from breed B, that is, full breed heterozygosity and full expression of heterosis. In a backcross mating a breed A sire to an F_1 dam, all genes from the sire originate from breed A, while on average, half the genes from the dam originate from breed A, and half originate from breed B. Therefore, it is expected that breed heterozygosity occurs at half the loci, and heterosis expression is halved. In an F_2 cross produced by mating F_1 sires and dams, half the genes from both sire and dam are derived from breed A, and half from breed B. On average, breed heterozygosity occurs at half the loci, and heterosis expression is again halved. In a 3-way cross derived by mating sires of a third breed (C) to F_1 dams, breed heterozygosity occurs at all loci, and heterosis is expressed fully. Maternal heterosis may be similarly explained, by examining the breed composition of the dam. For example, there is no maternal heterosis in the F_1 cross because the dams are breed B individuals. The backcross, the F_2 cross and the 3-way cross however all benefit from full maternal heterosis, because in each case the dams are F_1 individuals.

Using these concepts, the weighting approach can be extended to predict heterosis effects in different crosses. Firstly, a dominance effect is defined for each combination of breeds, (e.g. d_{AB} for the dominance effect between breeds A and B. These effects are weighted by the expression of heterosis in each crossbred or straightbred type, e.g. 0 for breeds A and B, 1 for the F_1, and $\frac{1}{2}$ for the F_2. These weightings are shown for a variety of straightbred and crossbred types in Table 12.1.

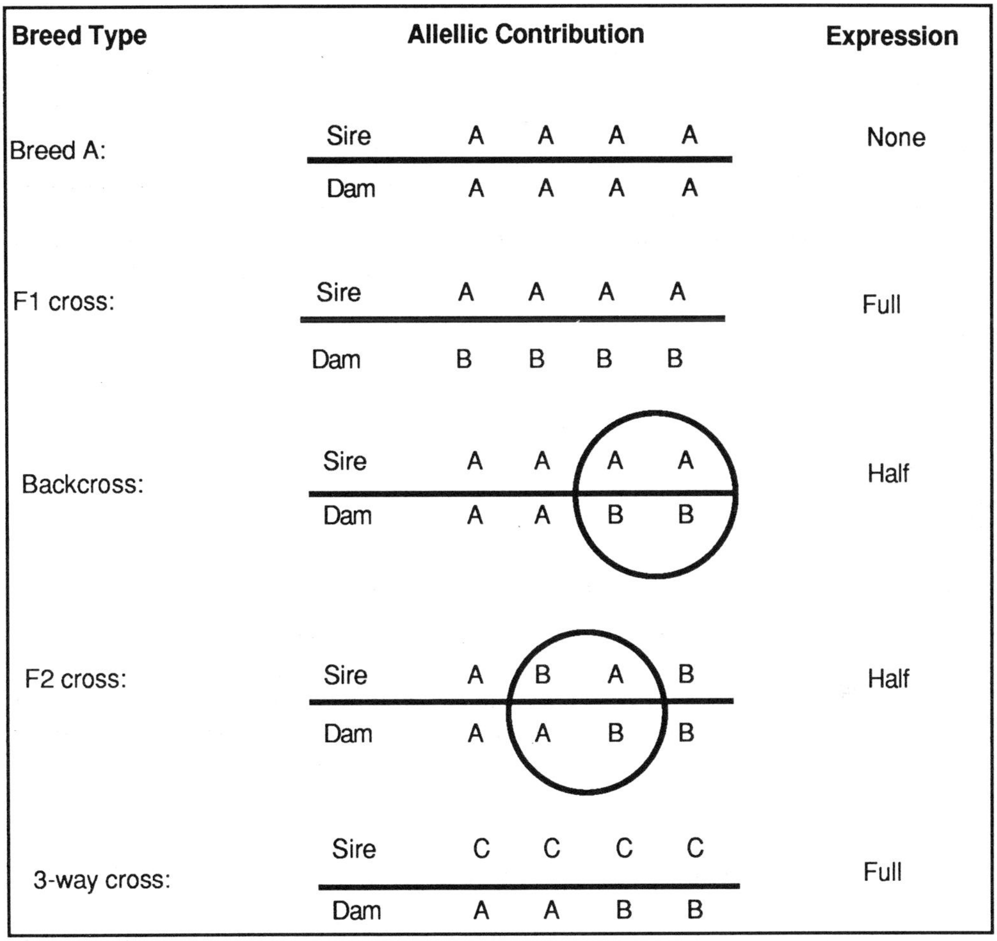

Breed Type	Allellic Contribution					Expression
Breed A:	Sire	A	A	A	A	None
	Dam	A	A	A	A	
F1 cross:	Sire	A	A	A	A	Full
	Dam	B	B	B	B	
Backcross:	Sire	A	A	A	A	Half
	Dam	A	A	B	B	
F2 cross:	Sire	A	B	A	B	Half
	Dam	A	A	B	B	
3-way cross:	Sire	C	C	C	C	Full
	Dam	A	A	B	B	

Fig. 12.1 Expression of heterosis in different straightbred and crossbred types

Table 12.1 Weightings on breed and heterosis effects (dominance). Maternal heterosis between breeds A and B is denoted as $d_{AB}{}^M$. In the F_1 crosses AxB uses breed B as the dam, and BxA uses breed A as the dam.

Straightbred or crossbred type	g_A	g_B	m_A	m_B	d_{AB}	$d_{AB}{}^M$
Breed A	1	0	1	0	0	0
Breed B	0	1	0	1	0	0
F_1 (AxB)	$\frac{1}{2}$	$\frac{1}{2}$	0	1	1	0
F_1 (BxA)	$\frac{1}{2}$	$\frac{1}{2}$	1	0	1	0
F_2	$\frac{1}{2}$	$\frac{1}{2}$	$\frac{1}{2}$	$\frac{1}{2}$	$\frac{1}{2}$	1
Backcross to A	$\frac{3}{4}$	$\frac{1}{4}$	$\frac{1}{2}$	$\frac{1}{2}$	$\frac{1}{2}$	1
Backcross to B	$\frac{1}{4}$	$\frac{3}{4}$	$\frac{1}{2}$	$\frac{1}{2}$	$\frac{1}{2}$	1

An important benefit of the weighting approach is that the performance of untested crossbred types can be predicted using the estimates of breed effects and heterosis. For example (Dickerson, 1969), the predicted performance of a 3-way cross using breed C terminal sires over F_1 (AxB) dams is:

$$\tfrac{1}{2}g_C + \tfrac{1}{4}g_A + \tfrac{1}{4}g_B + \tfrac{1}{2}m_A + \tfrac{1}{2}m_B + \tfrac{1}{2}d_{CA} + \tfrac{1}{2}d_{CB} + d_{AB}{}^M$$

where the subscript 'C' indicates breed C.

Simple Across-Breed Genetic Evaluations

Breed effects, heterosis, and within-breed Estimated Breeding Values (EBVs) can be combined into a Multibreed Selection Index (MSI: Kinghorn, 1986). Use of the MSI enables simultaneous exploitation of crossbreeding and selection. A relatively simple method of constructing an MSI, with application to the North American beef industry (Notter, 1989) is the Across-Breed EPD (Estimated Progeny Difference = $\tfrac{1}{2}$ Estimated Breeding Value), constructed as follows:

Across-breed EPD = Breed Effects + Heterosis + Base Adjustment + EPD

The estimates of breed effects and heterosis can be obtained from crossbreeding experiments. Because these components are included, it should be noted that the Across-Breed EPD depends on the breed type of prospective mates, i.e. an animal will have Across-Breed EPDs for matings with all breeds and crosses, for each trait evaluated.

The predictions of breed effects and heterosis required can be derived using the weighting approach described above. For this purpose, breed tables containing standard sets of estimates of breed effects and heterosis could be published. This would allow users to calculate Across-breed EPDs themselves, rather than publishing EPDs for each animal in each mating type.

The base adjustment is included to adjust animals of different breeds to the same zero EPD point, implicitly defined as the average of the base animals in the BLUP analysis. Base animals are the first animals appearing in the pedigree. If all breeds in an industry were to adopt the same zero EPD point in their BLUP analyses, the need to calculate base adjustment factors would be eliminated.

As mentioned above EPDs are essentially the same as EBVs, both being derived from BLUP models under identical assumptions. These are frequently published by breed organisations in sire summaries from across herd BLUP analyses. Consequently, the necessary information to construct Across-Breed EPDs is available to producers. To make the task simpler, breed tables containing appropriate estimates of breed effects and heterosis could be published, perhaps by research organisations. Also, it would be beneficial for all breeds to adopt the same zero point in their BLUP analyses.

The advantages of such an approach are:

- Animals of different breeds and crosses can be directly compared.
- It would enable use of genetic differences within and between breeds and crosses to more efficiently meet market requirements.

There are however several potentially serious drawbacks to this simple approach:

- It requires reliable estimates of breed effects and heterosis for all economically important traits and for all breeds and crosses in use, otherwise some breeds and crosses could be incorrectly disadvantaged.
- One set of reliable estimates for an industry assumes no breed or cross by environment interaction, which is an untenable assumption.
- The users of genetic evaluations, seedstock breeders and their clients, could be thrown into confusion with the added complexity of across-breed evaluations over within-breed evaluations, to the detriment of the former. Therefore, reporting systems for these across-breed evaluations should be designed thoughtfully. Extension programs will play a major role in explaining the reporting systems, and use of the results. These considerations also apply to the more complex across-breed evaluations described in the next section.

More Complex Evaluation Procedures

In some situations, more complex across-breed evaluation methods may be warranted. These methods involve including data from all breeds and crosses in a single BLUP analysis. One such situation is where all animals, straightbred and crossbred, are candidates for selection in a crossbreeding program. The across-breed EPD method above can only be used on straightbred animals, since these are the only animals currently analysed in genetic evaluations.

The simplest approach is to extend the within breed animal model to include breed effects and heterosis, as shown in Figure 12.2. This combines two models: the within-breed BLUP model described in Chapter 5, and the weighting approach for estimating crossbreeding effects described above. It is also possible to account for differences in genetic and environmental variances across straight bred and crossbred types, although not all procedures available are entirely appropriate.

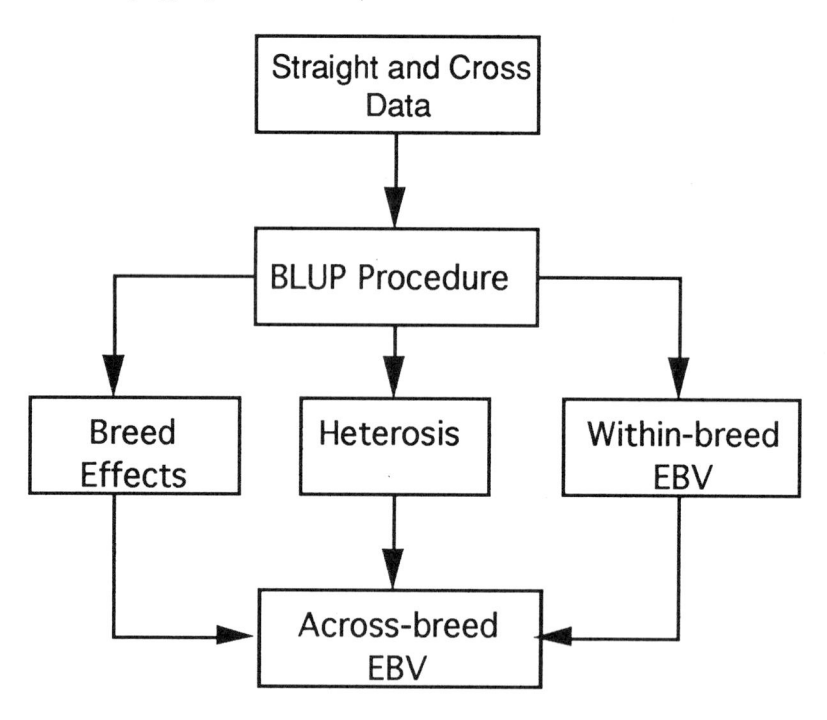

Fig. 12.2 Components of across-breed genetic evaluations

The estimates of breed effects, heterosis, and within-breed breeding value obtained from such a model can then be used to construct across-breed EBVs in the same way across-Breed EPDs were constructed above. The difference between the two approaches is that all components are now derived from the same analysis model, which is more desirable.

A potentially important problem not accounted for by the methods above is the effect of animal by breed-of-mate interactions on across-breed genetic evaluations. These effects can alter the expression of an animal's breeding value when mated to different straightbred and crossbred types. Specifically, the true breeding value rankings of a group of animals may change for different mating types. Consequently breeding values estimated from within-breed data may not be good indicators of performance in crossbred matings. The genetic correlation between straightbred and crossbred performance (r_{gsc}) can be used as an indicator of the importance of these effects. If r_{gsc} is close to 1, re-rankings are unimportant, and the evaluation methods described above may be adequate. If r_{gsc} is significantly less than 1, re-rankings may be important, and an alternative across-breed evaluation model may be required.

The biological model most appropriate is a multi-trait model, fitting each straightbred and crossbred type as a separate trait (Swan and Kinghorn, 1992). This should not be confused with the multi-trait model described in Chapter 5, in which traits are different characters, such as birth weight and weaning weight. In this case, each straightbred and crossbred type comprises a separate trait for a single character, say birth weight. This poses a problem for genetic evaluation systems: superimposing a number of straightbred and crossbred traits on a model already including several different characters will lead to very large analyses. This is not an insurmountable problem, as there are various computational tricks which can be used, and the rapid development of computers enables the use of more and more complicated models each year. A more serious problem is that selection on crossbred performance actually changes the value of r_{gsc}, and multi-trait models are not robust under these circumstances.

An alternative approach is as follows. Firstly, define a single crossbred type as the target for genetic improvement, e.g. a 3-way terminal cross. Then fit a single trait model to data from the target genotype only. By including all available pedigree information, breeding values for the target will be obtained for all animals used to produce it. Such an approach avoids the problems of the multi-trait model: computational demands are lower, and a single trait model is unaffected by changes in the value of r_{gsc}.

Before adopting such a genetic evaluation system, it is important to know the value of r_{gsc} for various economically important characters. Values reported in the research literature are mostly unreliable, coming from small data sets. To estimate r_{gsc} accurately, large well designed experiments are needed. It is likely that in characters showing small amounts of heterosis, r_{gsc} will approach 1, while for characters showing high levels of heterosis, r_{gsc} will be significantly lower than 1. These trends could be used as a guide to choose which characters to measure in experiments.

Industry Application of Across-Breed Evaluation

Adoption of across-breed evaluation systems will be strongly influenced by the structure of the industry involved and the importance of crossbreeding. This will now be discussed for the major livestock industries.

Beef cattle. The benefits of crossbreeding are well demonstrated by research in many beef industries, and the uptake of crossbreeding is increasing. However, there is currently little opportunity for seedstock breeders to collect data on crossbred animals. Genetic evaluation systems, e.g. BREEDPLAN in Australia, are used by breeders with purebred herds, mainly within breed organisations. Crossbred animals are bred by commercial producers who buy bulls from the seedstock breeders. It is unlikely that seedstock breeders will be able to obtain useful data from their commercial crossbreeding clients. The simple across-breed EPD method described above could be used by the beef industry with its current structure. However, this method may not be entirely suitable, as there is likely to be opposition from breed organisations who perceive their breeds to be disadvantaged.

The more complex across-breed evaluation systems described above are more appropriate in this regard. There is likely to be demand from producers operating crossbreeding programs, especially those involved in developing composite breeds or marketing crossbred seedstock. These producers have the ability to collect data on both straightbred and crossbred animals, which could be analysed using appropriate BLUP models. As the use of crossbreeding increases, the demand for superior crossbred animals will increase. This may cause changes to industry structures such that producers marketing crossbred seedstock play an important role. This is likely to be the first application of across-breed evaluations in the beef industry. Breed organisations need not be threatened by such developments as they have the opportunity to promote their breeds for specific roles in crossbreeding programs.

Dairy cattle. Dairy industries are commonly characterised by the widespread use of herd recording principally for management purposes, and AI. This has facilitated the use of advanced across-herd genetic evaluation and breeding systems. The Holstein-Friesian breed is currently dominant. If at some time in the future crossbred cows were to show an economic advantage, perhaps due to a change in the basis of payments for milk, or for sustainability arguments, it would be possible for the dairy industry to adopt across-breed evaluations including crossbred animals in analyses. The Australian dairy industry has recently introduced an across-breed evaluation to cover all red breeds, employing the procedures outlined in the previous section.

Sheep. The pre-eminance of the Merino in fine apparel wool industries means that crossbreeding in wool enterprises has little or no importance. Crossing different Merino strains may have some value, provided that there is no heterosis for fibre diameter, but is not widely promoted. Also, the Merino industry has only recently started using BLUP for genetic evaluations. Consequently, there is no potential for across-breed genetic evaluation in the wool industry in the near future.

Crossbreeding is used widely in meat sheep industries. For example, in Australia, a 3-way crossing system is almost universal, with F_1 Border Leicester x Merino females joined to a terminal sire breed such as the Dorset. BLUP is used for genetic evaluations in the terminal sire breeds, but information is not available from the dam side. The F_1 dams are produced by mating aged Merino ewes from the wool industry to Border Leicester rams. However, terminal sire breeders could collect information from crossbred progeny for an across-breed evaluation.

Pigs. The application of across-breed genetic evaluations has the greatest immediate potential in the pig industry. Crossbreeding is widespread internationally, with extensive use of a 3-way crossing system with sires from a high producing terminal line and F_1 dams derived by crossing maternal lines. In Australia, a within-breed genetic evaluation and indexing system (PIGBLUP) is available for improvement of straightbred seedstock. Other countries also use sophisticated within-breed evaluation procedures. The structure of the industry internationally makes it amenable to across-breed evaluations. Performance recording is important, and in some cases the seedstock and crossbreeding segments are contained in the same production unit, so information on crossbred animals is readily available. Also, because of the high reproductive rate of pigs, a large part of the population can be crossbred.

Application of an across-breed evaluation model including straightbred and crossbred data would be possible.

Guide to Consultants

Across-breed genetic evaluation is not yet commonly available for any livestock industry. Breeding consultants are unlikely to have to make specific recommendations on the subject for some time. However, there are considerable advantages in combining selection and crossbreeding strategies, so some general points can be made:

- The need for across-breed evaluations depends on the value of crossbreeding to the industry in question. Crossbreeding is beneficial in the beef cattle, pig, and meat sheep industries, while there is currently little value in crossbreeding in the dairy cattle and wool sheep industries.
- Several different across-breed procedures are available, from simply combining EBVs from within breed evaluations with results from crossbreeding experiments, to more comprehensive methods including data from both straightbred and crossbred animals in a single BLUP procedure.
- The adoption of a procedure depends largely on industry structures. For example, collection of data on crossbred animals is difficult in the beef industry, which may initially inhibit the uptake of across-breed evaluations. However, where such evaluations can be demonstrated to be beneficial, industry structures may change.
- Further research is needed in a variety of production systems to evaluate the importance of re-rankings of animals when mated to different breeds and crosses.
- Before implementing across-breed evaluation procedures, considerable effort must be made to develop data collection and reporting strategies which users can easily understand. Extension specialists will have an important role in this area.

References

Dickerson GE (1969) Experimental approaches in utilizing breed resources. Anim Breed Abstr 37:191-202

Kinghorn BP (1986) Mating plans for selection across breeds. 3rd World Conf Genet Appl Livest Prod Vol XII 233-244

Notter DR (1989) EPDs for use across breeds. Proc Beef Impr Fed, 21st Meeting, Nashville, TN, May 11-13, 1989, 63-78.

Swan AA and Kinghorn BP (1992) Evaluation and exploitation of crossbreeding in dairy cattle. J Dairy Sci 75:624-639.

PART III: Breeding Objectives

Chapter 13

Introducing Economics to Modern Animal Breeding

Stephen Barwick

Introducing the Breeding Objective

Animal Breeding was defined in Chapter 1 as the genetic manipulation of biological differences between animals over time using approaches aimed at maximising profitability. What is the direction of change which maximises profitability? What is the combination of differences, for traits, between animals which maximises returns over costs? How is this direction established? These are economic questions about the biology of production and product quality. They constitute what we term the **breeding objective**.

Maximising profitability will involve not just one output trait but several output and input traits. The need to establish the required balance of these traits has been appreciated increasingly over the past two decades. The breeding objective sets the breeding direction and provides the basis against which the effectiveness of the breeding program can be gauged. This chapter is concerned with the process of establishing this required balance of traits, and with the involvement of economics in this. There are, of course, also many other important applications of economics in animal breeding, including methods of valuing individuals and of evaluating alternative recording schemes and breeding programs. These are summarised in Chapters 23 and 24.

Some Important Distinctions

Selection and culling. Selection is concerned with whether an animal should be chosen to breed the next generation. **Culling** is concerned with whether an animal that has not been selected should be completely removed from the production herd or flock. Culling decisions take account of the likely performance of animals, usually in their next year or two, while selection is mostly concerned with the likely performance of the progeny of animals. Culling is said to be concerned with the current herd or flock, and selection mostly with the future herd or flock. Culling decisions utilise a knowledge of the **repeatabilities** of traits, while selection decisions utilise a knowledge of the **heritabilities** of traits and their genetic associations. Detailed evaluations can be performed for either and both purposes, i.e. to maximise profitability of the current herd or flock or to maximise longer term profitability. Strictly, genetic evaluations for selection might also take account of expected current herd or flock performance in some cases, since this also influences profitability.

Performance in the current and future herd or flock. The distinction between performance in the current herd or flock and the future herd or flock can be seen for the trait ewe fleece weight. Ewes are clearly able to be ranked both for their own expected future fleece production, current flock performance, and for their breeding value for fleece weight, future flock performance. Other traits in which performance is measured repeatedly over the life of the animal include many reproductive traits. Traits are usually more repeatable than they are heritable, and that some reproductive traits, in particular, can be markedly less heritable than they are repeatable.

Traits of the breeding objective and selection criteria. A distinction needs to be made between the **traits** of the breeding objective, which are desired to be improved, and the available measures, or **selection criteria,** which are useful because of the information they provide about the traits. The traits are sometimes described as ends, and selection criteria as the means to those ends.

The traits for inclusion in the breeding objective should be decided by their importance irrespective of their ease or cost of measurement. Consequently, they will often differ from the available selection criteria, depending on whether or not the traits themselves are able to be readily and inexpensively measured. Selection scenarios encountered can therefore range from simple mass selection, direct selection based on a single phenotypic record on each individual, through simple indirect selection and various combinations of direct and indirect selection. Some of these scenarios are shown, schematically, below. An example of a measure that is a selection criterion rather than a trait is perhaps scrotal size, because it does not have economic importance in its own right, but rather has importance because of the information it conveys about male and female fertility.

Illustration of some types of selection scenarios encountered

Available selection criteria		Traits to be improved
A	→	A
B	→	A
A, D, G	→	D, F
L, P, Q	→	A, D, F, G

Selection for more than one trait

An obvious method of selection to improve more than one trait is to select for one trait for a period of time, then for another single trait for a period, and so on. This is **tandem selection,** and it is usually the least efficient method when the real goal is to improve several traits. The next alternative is to set levels for each of several traits below which animals will not be selected - that referred to in most texts as the method of **independent culling levels.** The independent culling level method is intermediate in efficiency between tandem selection and **index selection,** the latter being where the most appropriate weightings are determined for each trait and then applied in a linear index. Each method is described here as though the traits themselves are measured, but as mentioned, this is not often the case and the methods can be thought of as being applied to whatever selection criteria are available.

Two examples of the development of simple selection indexes follow. A brief background to these procedures is first given, for reference. Example 1 involves prediction of the breeding value of a single trait; weight at 480 days in beef cattle, while Example 2 involves prediction of a function of the breeding value for weight at 480 days and the breeding value for cow weaning rate.

Some Background to the Derivation of Selection Indices

The single-trait objective

Let us first consider the somewhat artificial situation where the objective is to improve the breeding value, A, for only one trait, and there are, say, 3 measurements (P_1, P_2, P_3) available for use as selection criteria. Then the most appropriate weightings (bs) to apply to each measurement are obtained by solving the following set of equations:

$$b_1 \, Cov(P_1,P_1) + b_2 \, Cov(P_1,P_2) + b_3 \, Cov(P_1,P_3) = Cov(P_1,A)$$

$$b_1 \, Cov(P_2,P_1) + b_2 \, Cov(P_2,P_2) + b_3 \, Cov(P_2,P_3) = Cov(P_2,A)$$

$$b_1 \, Cov(P_3,P_1) + b_2 \, Cov(P_3,P_2) + b_3 \, Cov(P_3,P_3) = Cov(P_3,A)$$

These are known as the selection index equations. They can also be written, in matrix notation, as

$$\mathbf{Pb = G}$$

P being a matrix of phenotypic variances and covariances among the measurements, **b** the vector of unknown weightings to be solved for, and **G** a matrix of genetic covariances between the measurements and the objective, A.

i.e.,

$$\begin{bmatrix} Cov(P_1,P_1) & Cov(P_1,P_2) & Cov(P_1,P_3) \\ Cov(P_2,P_1) & Cov(P_2,P_2) & Cov(P_2,P_3) \\ Cov(P_3,P_1) & Cov(P_3,P_2) & Cov(P_3,P_3) \end{bmatrix} \begin{bmatrix} b_1 \\ b_2 \\ b_3 \end{bmatrix} = \begin{bmatrix} Cov(P_1,A) \\ Cov(P_2,A) \\ Cov(P_3,A) \end{bmatrix}$$

$$= \begin{bmatrix} Cov(A_1,A) \\ Cov(A_2,A) \\ Cov(A_3,A) \end{bmatrix}$$

The resulting index of measurements and their weightings is of the form

$$I = \hat{\mu} + b_1(P_1 - \hat{\mu}_1) + b_2(P_2 - \hat{\mu}_2) + b_3(P_3 - \hat{\mu}_3)$$

where individual animal values for P_1, P_2 and P_3 are expressed as deviations from their respective group means, and $\hat{\mu}$ is estimated from the observed group mean for the objective trait. Index values calculated for each animal are predictions of A. Provided any necessary adjustments have been made for systematic sources of variation in the measurements, e.g. management group effects, these index values are the most accurate linear predictions of A possible from the available measurements.

Note that to set up the selection index equations requires a knowledge of a number of covariances. These can be obtained from estimates of phenotypic ($r_{P_1P_2}$) and genetic ($r_{A_1A_2}$) correlations, heritabilities (h^2) and phenotypic variances (σ_P^2, or V_P), since

$$r_{P_1P_2} = \frac{Cov(P_1,P_2)}{\sigma_{P_1}\sigma_{P_2}}$$

$$r_{A_1A_2} = \frac{Cov(A_1,A_2)}{\sigma_{A_1}\sigma_{A_2}}$$

$$\text{and } h^2 = \frac{V_A}{V_P}$$

Each measurement can be expressed as a sum of genetic and environmental effects, which in the simplest case might involve only additive genetic effects, giving rise to the breeding value A, and an environmental effect $\mathcal{E}$ peculiar to the measurement made on each animal, i.e.

$$P_1 = A_1 + \mathcal{E}_1.$$

Genetic and environmental effects are assumed to be uncorrelated.

Also needed is the fact that the covariance between linear functions of variables can be expanded much like other arithmetic expressions. Where A, B, C and D are variables, and p, q, f and g are any constants,

$$Cov(pA + qB, fC + gD) = pf\,Cov(A,C) + pg\,Cov(A,D)$$
$$+ qf\,Cov(B,C) + qg\,Cov(B,D)$$

Also, $Cov(A,A) = Var(A)$

A variance is thus just a special type of covariance, which derives from the fundamental definitions of each, in terms of mathematical expectations (E), viz.

$$Cov(X,Y) = E[XY] - E[X]\,E[Y]$$

$$Var(X) = E[X^2] - (E[X])^2$$

Note, further, that while

$$Cov(P_1,P_2) = Cov(A_1 + \varepsilon_1, A_2 + \varepsilon_2)$$

$$= Cov(A_1,A_2) + Cov(\varepsilon_1, \varepsilon_2)$$

when P_1 and P_2 are measured on the same animal,

$$Cov(P_1,P_2) = a\,Cov(A_1,A_2)$$

if P_1 and P_2 are measured on different animals, where a is the coefficient of relationship between the animals. Common values of a include 0.5 between sire or dam and offspring, 0.5 between full sibs, and 0.25 between half sibs.

The multi-trait objective

For the more common situation where there is more than one trait in the breeding objective, the selection index equations become

$$\mathbf{Pb} = \mathbf{Gv}$$

where **P** and **b** are as before and **Gv** is

$$\begin{bmatrix} Cov(A_1,A_{T_1}) & Cov(A_1,A_{T_2}) \\ Cov(A_2,A_{T_1}) & Cov(A_2,A_{T_2}) \\ Cov(A_3,A_{T_1}) & Cov(A_3,A_{T_2}) \end{bmatrix} \begin{bmatrix} v_{T_1} \\ v_{T_2} \end{bmatrix}$$

for an example case where there are two traits (T_1 and T_2) in the objective. Here, the values in **v** are usually economic values, attributable to each of the traits in the objective.

Selection Index Examples

Example 1: One trait in the objective, two selection criteria available

Objective: Prediction of the breeding value, A, for each animal, for weight at 480 days in beef cattle

Measures available: A single measurement, P_1, of 400 day weight on each animal, and a single measurement, P_2, of scrotal circumference on the *sire* of each animal

Other estimates:

Selection criteria

σ_{P_1} = 32.6 kg
σ_{P_2} = 2.06 cm
h_1^2 = 0.30
h_2^2 = 0.42
$r_{P_1 P_2}$ = 0.40
$\hat{\mu}_1$ = 400 kg
$\hat{\mu}_2$ = 29 cm

Objective traits

σ_P = 32.6 kg
h^2 = 0.30
$\hat{\mu}$ = 420 kg

Between criteria and trait

$r_{A_1 A}$ = 0.80
$r_{A_2 A}$ = 0.30

The problem is to define an index of the form

$$I = 420 + b_1(P_1 - 400) + b_2(P_2 - 29),$$

and the issue is to find the values of b_1 and b_2 which are most appropriate.

Setting up the equations,

$\text{Cov}(P_1, P_1) = V_{P_1} = (32.6)^2 = 1062.76$

$\text{Cov}(P_1, P_2) = \text{Cov}(P_2, P_1) = 1/2 \cdot r_{P_1 P_2} \cdot \sigma_{P_1} \cdot \sigma_{P_2} = 1/2 \, (0.40)\,(32.6)\,(2.06)$
$\phantom{\text{Cov}(P_1, P_2) = \text{Cov}(P_2, P_1) = 1/2 \cdot r_{P_1 P_2} \cdot \sigma_{P_1} \cdot \sigma_{P_2}} = 13.431$

(Note: 1/2 because of the relationship between sire and offspring)

$\text{Cov}(P_2, P_2) = (2.06)^2 = 4.2436$

$$\text{Cov}(P_1, A) = \text{Cov}(A_1 + \mathcal{E}_1, A)$$
$$= \text{Cov}(A_1, A)$$
$$= r_{A_1 A} \cdot \sigma_{A_1} \cdot \sigma_A$$
$$= r_{A_1 A} \cdot h_1 \sigma_{P_1} \cdot h \sigma_P$$
$$= (.80)\sqrt{0.30}\,(32.6)\sqrt{0.30}\,(32.6)$$
$$= 255.06$$

$$\text{Cov}(P_2, A) = \text{Cov}(A_2 + \mathcal{E}_2, A)$$
$$= 1/2 \cdot r_{A_2 A} \cdot h_2 \sigma_{P_2} \cdot h \sigma_P$$
$$= 1/2\,(0.30)\sqrt{0.42}\,(2.06)\sqrt{0.30}\,(32.6)$$
$$= 3.5757$$

Hence the equations are

$$b_1(1062.76) + b_2(13.431) = 255.06$$
$$b_1(13.431) + b_2(4.2436) = 3.5757$$

or in matrix notation,

$$\begin{bmatrix} 1062.8 & 13.431 \\ 13.431 & 4.2436 \end{bmatrix} \begin{bmatrix} b_1 \\ b_2 \end{bmatrix} = \begin{bmatrix} 255.06 \\ 3.5757 \end{bmatrix}$$

which on solving gives

$$b_1 = 0.2389$$
$$b_2 = 0.0865$$

The index which best predicts A from the information available is therefore

$$I = 420 + 0.2389\,(P_1 - 400) + 0.0865\,(P_2 - 29)$$

The accuracy of this index, given by σ_I/σ_A, is 0.44 since

$$\sigma_I^2 = (0.2389)^2 \cdot \sigma_{P_1}^2 + (0.0864)^2 \cdot \sigma_{P_2}^2 + 2\,(0.2389)\,(.0864)\,\text{Cov}(P_1, P_2)$$
$$= 61.24,$$
$$\sigma_I = 7.83,$$

and $\sigma_A = h\sigma_P = \sqrt{0.30}\,(32.6) = 17.85$

Suppose now that the records available on three animals X,Y and Z, say, are:

	P_1	P_2
Animal X:	383	33.0
Animal Y:	410	29.0
Animal Z:	386	26.0

The index values for X, Y and Z are then 416.0, 422.4 and 416.2, respectively. These are the predicted breeding values, $\hat{A}$, for each animal. The ranking of animals for $\hat{A}$ is consequently Y then Z then X.

Example 2: Two traits in the objective, two selection criteria available

Objective: Prediction of the breeding value for weight at 480 days (WT) and the breeding value for cow weaning rate (WR) in beef cattle, assuming that we know that a *unit* of weaning rate is, say, four times as valuable as a unit of weight.

i.e., the objective is $A_{WT} + 4A_{WR}$,

and we will call this H.

Measures available: P_1 and P_2, as for Example 1.

Other estimates: Selection criteria   Objective traits

as for Example 1

$\sigma_{P_{WT}}$ = 32.6 kg
h^2_{WT} = 0.30
$\hat{\mu}_{WT}$ = 420 kg
$\sigma_{P_{WR}}$ = 39 calves
h^2_{WR} = 0.05
$\hat{\mu}_{WR}$ = 70 calves per 100 cows

between criteria and traits

$r_{A_1 A_{WT}}$ = 0.80
$r_{A_2 A_{WT}}$ = 0.30
$r_{A_1 A_{WR}}$ = 0
$r_{A_2 A_{WR}}$ = 0.25
$r_{A_{WT} A_{WR}}$ = 0

Setting up the equations,

$$\begin{aligned}
\text{Cov}(P_1, H) &= \text{Cov}(P_1, A_{WT} + 4A_{WR}) \\
&= \text{Cov}(A_1 + \varepsilon_1, A_{WT} + 4A_{WR}) \\
&= \text{Cov}(A_1, A_{WT} + 4A_{WR}) \\
&= \text{Cov}(A_1, A_{WT}) + 4\text{Cov}(A_1, A_{WR}) \quad \text{(see Example 1)} \\
&= 255.06 + r_{A_1 A_{WR}} \cdot \sigma_{A_1} \cdot \sigma_{A_{WR}} \\
&= 255.06 + 0 \\
&= 255.06
\end{aligned}$$

$$\begin{aligned}
\text{Cov}(P_2, H) &= \text{Cov}(A_2 + \varepsilon_2, A_{WT} + 4A_{WR}) \\
&= \text{Cov}(A_2, A_{WT}) + 4\text{Cov}(A_2, A_{WR}) \\
&= 3.5757 + 4[1/2 \, r_{A_2 A_{WR}} \cdot \sigma_{A_2} \cdot \sigma_{A_{WR}}] \\
&= 3.5757 + 4[1.4553] \\
&= 9.3969
\end{aligned}$$

and note that the left-hand side is unchanged from Example 1.

Hence the equations can be written

$$\begin{bmatrix} 1062.8 & 13.431 \\ 13.431 & 4.2436 \end{bmatrix} \begin{bmatrix} b_1 \\ b_2 \end{bmatrix} = \begin{bmatrix} 255.06 \\ 9.3969 \end{bmatrix}$$

or, alternatively, the right-hand-side can be written

$$\begin{bmatrix} 255.06 & 0 \\ 3.5757 & 1.4553 \end{bmatrix} \begin{bmatrix} 1 \\ 4 \end{bmatrix}$$

for which the solutions are

$$b_1 = 0.2208$$
$$b_2 = 1.5154$$

The resulting index is

$$I = \mu + b_1(P_1 - 400) + b_2(P_2 - 29),$$

or $\quad I - \mu = b_1(P_1 - 400) + b_2(P_2 - 29)$

$\quad\quad\quad = 0.2208(P_1 - 400) + 1.5154(P_2 - 29)$

if index values are expressed as deviations (i.e. as $I - \mu$).

If animals X, Y and Z have the same records on P_1 and P_2 as were given in Example 1, then their index values, as deviations, now are +2.3, +2.2 and -7.6, making X rank ahead of both Y and Z.

The **accuracy** of this index (σ_I/σ_H), is 0.21. Note that if records were available only for sire scrotal circumference, P_2 and not for 400 day weight, the b value to apply to P_2 obtained similarly, but by solving a new set of equations, would be 2.2144, and the accuracy of this new index for the same objective is 0.12.

The **relative efficiency** of the above two indexes, that with both weight and scrotal circumference measures and that using scrotal circumference alone, is given by the ratio of their accuracies, which indicates that for the objective $A_{WT} + 4A_{WR}$, the index based only on sire scrotal circumference is 57% as efficient as the index including both weight and scrotal circumference measures.

A further point is that the values 4 and 1, used to weight weaning rate and weight in this example, could equally well have been 8 and 2 or other numbers in the same ratio. It is only the *relative* magnitude of these values that affects the breeding direction, and the proportionality of differences between individual animals remains unaltered provided the relativity of these values is maintained. The values here were of course arbitrarily chosen, and a better procedure is to utilise estimates of the relative economic importance of each trait. Such economic estimates are the focus of the remainder of this chapter.

Selection for Compound Traits

Multi-trait problems are sometimes formulated as single trait problems by compounding traits into an apparently single trait. An example is the meat-sheep trait 'total weight of lean tissue per ewe joined' (TWL/EJ). This is really the product or compound of numerous traits, *viz.*

ewe fertility (ewes lambing per ewe joined) (EL/EJ)
•
ewe fecundity (lambs born per ewe lambing) (LB/EL)
•
lamb survival rate (lambs sold per lamb born) (LS/LB)
•
average lamb liveweight (LW)
•
average dressing percentage (CW/LW)
•
average % lean (of carcase weight) (WL/CW)

i.e. $$\frac{TWL}{EJ} = \frac{EL}{EJ} \cdot \frac{LB}{EL} \cdot \frac{LS}{LB} \cdot LW \cdot \frac{CW}{LW} \cdot \frac{WL}{CW}$$

Such products seem to avoid the difficulty of calculating economic values for traits. In addition, they are sometimes suggested as being desirable because they are **biological indexes**. However, while they do partly avoid the calculation of economic values, each component trait still has an **implied**, uncontrolled, **value** when selection is for the compound trait. Variation in the compound trait is in fact usually dominated by the most variable component trait. Since the compound trait can be improved via change in any component, and each component may contribute with differing cost effectiveness, it follows that it is not efficient to rely on the values that are implied in the compound trait.

A closely allied situation is that of selection for a **ratio** of traits. Cattle breeders, for example, might be interested in some ratio of eye muscle area and liveweight. A ratio is a form of product, so similar difficulties apply. The traits in the ratio have implied, uncontrolled values, and what is achieved by selection on the ratio depends very much on the relative genetic and phenotypic variabilities of the numerator and denominator traits.

Using Economics to Formulate the Breeding Objective

When there are a number of traits to be improved, the breeding objective can be expressed in the form

$$H = v_1 A_1 + v_2 A_2 + v_3 A_3 + ...$$

where the As are breeding values for the traits, and the vs are economic values reflecting the relative importance of each trait.

Definition of an economic value

A trait economic value is the marginal change in profit, or economic efficiency associated with a unit change in the trait, while other traits are held constant. For economic efficiency see definition below in the section describing effects of differing perspectives. Ideally, it is the partial derivative of a function describing profit, or economic efficiency, with respect to the trait in question. In practice, approximations to this may be used. A trait economic value should reflect the economics of production in the commercial sector of the industry, rather than in the seedstock sector, as it is the well-being of the commercial sector that is vital to all others for survival in the medium and longer term.

In the previous example, values of 4 and 1 were applied to the traits cow weaning rate and 480 day weight in beef cattle. Example estimates of economic values for these traits are $2.27 per percent and $0.71 per kg of normal sale animal, on a per cow basis, respectively. Notice, for this example, that the relative values are not that different from 4 and 1. The objective in Example 2 might thus have been written,

$$H^* = 0.71 A_{WT} + 2.27 A_{WR}$$

The appropriate predictor of H^* is the following index, which when expressed as a deviation is

$$I^* = 0.1594(P_1 - 400) + 0.8723(P_2 - 29)$$

for the same measures as previously. The accuracy of this index is 0.25. It can be shown that little is lost in the efficiency of selection for H* from using the two-measurement index from Example 2 rather than I*, illustrating the robustness of the procedures to minor variation in the estimated economic values.

Describing the relativity of economic values

Trait economic values refer to the value of an observed, or phenotypic unit of each trait. A unit, however, represents more of the total variation in some traits than in others. To meaningfully describe the relativity of trait economic values, i.e. in isolation from the selection index equations, requires that they be expressed in some standard amount of each trait's variability. The amount usually chosen for this is one phenotypic standard deviation (σ_P) of each trait.

In beef cattle, ball-park relative economic values for reproduction, growth and carcase traits are sometimes quoted as being 10: 2: 1. These are for a phenotypic standard deviation of each trait. In practice, important variation occurs around these figures.

A measure which goes further than the usual concept of trait economic values, by providing an indication of the relative importance of traits on the genetic scale, is the value of one *genetic* standard deviation (σ_A) of each trait. Another measure of relative *genetic* importance is the product of the economic value and the trait heritability; this yields smaller values (by a factor h/σ_P) than that for a genetic standard deviation.

The measurement basis for economic values

Special attention needs to be paid to units of measurement when deriving trait economic values and when appraising values calculated by others. Trait economic values can differ in their relativities, as well as in their absolute size, simply as a consequence of their basis of calculation. Apparent differences between studies can thus exist for this reason alone. Values might be calculated from the change in profit for a whole enterprise, in which case they will depend on the number of breeding females in the enterprise, or they might be calculated per breeding female, per individual sale animal or per unit of output from the enterprise. They can also differ if discounting has been used, as this introduces a variable time horizon over which the discounting may have been carried out.

Trait economic values may also be expressed on some basis, i.e. enterprise, cow, individual or unit of output, other than that on which they were calculated. Provided values are calculated on the same basis, their expression in different forms will affect the absolute, but not the relative sizes of trait economic values.

Steps in defining the breeding objective

The process of defining the objective can usefully be considered in four steps. These have been described by Ponzoni for several species and a recent example is given by Ponzoni and Gifford (1990) for cashmere goats. The steps are:

- **Identification of the breeding, production and marketing system**

  This involves such things as whether the animals are to have a purebreeding or crossing role; and characteristics of the commercial production unit and its product, including details of the herd structure, replacement and other management policies, ages of turnoff of slaughter animals and market requirements. Any prevailing restrictions on production or marketing also need to be identified.

- **Identification of sources of returns (R) and costs (C) in commercial herds or flocks**

  This allows development of a function describing profit (P), or economic efficiency, for commercial production,

  i.e., $P = R - C$

  where expansions of R and C describe component returns and costs. For the cashmere goat example of Ponzoni and Gifford,

  R includes:

(amount of fibre)	•	(value of a unit of fibre) for young goats
(amount of fibre)	•	(value of a unit of fibre) for breeding does
(number sold)	•	(value of each) for surplus offspring
(number sold)	•	(value of each) for cull does

  and C includes:

(amount of feed eaten)	•	(cost per kg of feed) for young goats
(amount of feed eaten)	•	(cost per kg of feed) for breeding does husbandry costs for kids, young goats and breeding does
(amount of fibre produced)	•	(cost of harvesting and marketing a unit of fibre) for young goats
(amount of fibre produced)	•	(cost of harvesting and marketing a unit of fibre) for breeding does
(number of animals sold)	•	(cost of marketing each) for surplus offspring
(number of animals sold)	•	(cost of marketing each) for cull-for-age does and fixed costs.

  Fixed costs are those which are independent of the level of production.

- **Determination of the biological traits influencing returns and costs**
  The profit equation is now expressed as a function of the principal biological traits contributing to each source of returns and costs. For the cashmere goat example, the traits identified by Ponzoni and Gifford were

  > down weight in young goats
  > down weight in breeding does
  > down fibre diameter in young goats
  > down fibre diameter in breeding does
  > number of kids weaned
  > liveweight in young goats
  > liveweight in breeding does
  > feed intake in young goats
  > and feed intake in breeding does

- **Derivation of the economic value of each trait**
  The change in profit, or economic efficiency, resulting from a unit change in each trait is determined assuming all other traits remain constant. In the example of Ponzoni and Gifford, there is the added constraint that overall stocking pressure remain constant. Profit is described by

  $$P = \Sigma_i \text{ factor}_i (R_i - C_i) X_i - K,$$

  where R_i and C_i are the returns and costs for the trait X_i, factor_i is the relevant discounted number of expressions of traits over the assumed time horizon, and K are fixed costs. P is evaluated at existing mean values for all traits, and then again (P^*) with the trait in question incremented by one unit. $P^* - P$ gives the trait economic value.

Issues in the Derivation of Trait Economic Values

The level of detail attempted

This applies both to the specification of the traits in the objective and to the extent of itemisation of returns and costs. As we saw earlier, it can make a difference to the objective if traits are expressed in composite, eg. product form rather than as components. Similarly, the detail with which returns and costs are itemised can affect economic values. While it is obviously best to be as thorough as possible in any formulation, this needs to be tempered by a realisation that the economic values obtained can only ever be estimates. It is also to be expected that individuals will differ in their interpreration of the exact formulation of traits to include in any objective.

It is generally not critical that exact trait economic values are obtained, as the efficiency of the selection index is quite robust to modest changes. It is important that the estimates of the economic values are of correct sign and relative order of magnitude.

Whose economics to consider?

Economic values should be assessed at the level of the commercial herd or flock. Having said this, there are instances where this issue is not absolutely clear-cut, and where debate continues. These cases, however, are generally of lesser importance and represent refinements to the overall principle. Perhaps the most common example where a question arises is in the case of a terminal-sire line, since genes for female reproduction have no value to the commercial or crossbreeder. If completely ignored in the terminal-sire objective, however, profitability of the seedstock breeder can ultimately be affected through there being less seedstock animals to sell. Other cases can also be visualised where more than one set of economic signals could influence the objective, and in these cases a sensible approach to use may be to weight the values for each situation by the estimated proportion of profit associated with each.

Estimation methods

Budgeting methods are usually used, with a complete budget for the enterprise being best, allowing construction of an equation describing profit or economic efficiency. The economic value is then the partial derivative of profit, or economic efficiency, with respect to the trait in question. Alternatively, partial budgeting can be used, necessitating only that the changes in returns and costs, associated with changes in traits, are detailed. Similar results are obtained with partial budgeting provided the unit of change in each trait is small. When relationships between traits and production factors are too complex to be adequately described by a single profit equation, simulation utilising a bio-economic model may be preferred.

Effects of differing perspectives and management and marketing constraints

We saw earlier that it is the **relativity** of trait economic values that sets the breeding direction. This relativity can be affected by the perspective taken in estimating trait economic values. If the perspective is a national one, or that of a whole industry, it is best that the trait economic values do not depend on the scale of the assumed enterprise, in which case the appropriate values are those computed **per unit of product.** Values computed per unit of product are the same as those computed per enterprise, or per breeding female, where there is the added constraint of fixed total output, so that if an individual commercial producer operates under a production quota, values computed per unit of product are also appropriate. Similar results are obtained if total inputs are fixed rather than total output. Economic values computed per unit of product are the same as those computed by combining the components of profit (R and C) as a ratio rather than as a margin (R - C), the ratio being another form of constrained profit which is referred to as **economic efficiency.**

A common perspective is that of an individual commercial producer operating in the absence of production quotas in a free-enterprise environment. For this perspective, it is usual for trait economic values to be calculated from the associated change in **unconstrained profit.** This is based on the usual economic assumption that producers are profit maximisers. If values are instead calculated from changes in economic efficiency, i.e. the ratio of costs and returns, the assumption being made is that producers prefer to minimise costs rather than to maximise profit. While this may be true in some cases, it is probably more likely that restrictions will exist on particular cost inputs such as the availability of feed, rather than on total inputs, in which case it is only necessary to take account of the particular restrictions for the trait values that are affected.

Non-linearity of economic values

When the economic value for a trait depends on the level of expression of the trait, the economic value is non-linear. An example is litter size at birth in sheep; the value of an extra lamb born decreases as the number in the litter increases, there being extra management requirements in achieving a similar level of output for each additional lamb. Provided the present level of performance in such traits is taken as the base, and the unit of change is small, a linear approximation is usually an adequate approximation of the economic value.

The need for optimal management

Calculations of trait economic values should be made assuming that the management prevailing is optimal. This applies both to the base situation, before improvement, and to the situation after change in a trait. If the improvement necessitates a change in management, this change should be assumed to have been made when assessing the economic value. Alterations to management will usually be small if the unit of change being considered in the trait is small.

An example trait where the assumed management can have an effect on the assessed economic value is cow fertility, where it is common to ignore the fact that dry cows are best sold as soon as possible after being diagnosed empty. Hence, as well as the additional income from an extra calving cow, the lost income from sale of one less dry cows has to be taken into account. Of course, there are also other effects on profit to consider that arise from the altered herd structure which results from culling fewer dry cows.

Accounting for feed costs

Feed costs cannot be ignored because they are often the largest of all costs. Ideally, feed intake should be included as a trait in the breeding objective, but a shortage of genetic parameter estimates for this trait, and concern about the reliability of the estimates that are available, often preclude this approach. A less desirable approach sometimes used is to presume feed requirements are a direct function of liveweight. The costs of any additional feed needed are then incorporated as a cost in the economic values calculated for liveweight traits.

There are essentially two approaches used for costing feed. The first allows additional feed to be purchased, in some form, while the second requires that extra feed needed is offset by reducing stock numbers, usually breeding females. These approaches apply approximately to intensive and extensive production situations.

Accounting for competitive position

Methods exist (deVries, 1989) for taking account of the competitive position of the breeding unit, in relation to other breeders, in deriving trait economic values. Values are appropriately reduced for traits for which an advantage already exists, and *vice versa*.

Discounting

Traits differ in the time taken for any improvement made to be expressed in the herd or flock. Discounting procedures can be used to adjust economic values for this. Methods available include the **discounted gene flow** method (McClintock and Cunningham, 1974) and a more recently developed method of **diffusion coefficients** (McArthur and del Bosque Gonzalez, 1990). Gene flow accounts for both the time delay between selection and when descendents are born and for the delay between the birth of a descendent and the time when the decendent has an effect on cash flow. The diffusion coefficient method accounts only for the latter delay, the argument being that this is sufficient for estimation of trait economic values. For other applications, such as valuing individual animals, or valuing an infusion of a new genotype, it is likely to be more important to account for both types of time delays.

Other Ways of Selecting for More Than One Trait

Selection indices for improving more than one trait need not use economics as their basis, but instead may be constructed according to other constraints. Indices that utilise other constraints are known as **constrained or restricted indices.** The constraint may apply to a particular trait and be imposed jointly with other trait economic values, or constraints may be used to completely replace the need for economic values.

A common restricted index is one where change in one trait is desired to be held to zero while improvement is made in one or more other traits. An example could be the desire to hold fibre diameter constant in Merinos while increasing fleece weight. To achieve this, the constraint that is imposed is that the covariance between the index and the breeding value for fibre diameter is zero.

Another type of restricted index is the **desired gains** index, where the absolute amounts, or equally, the ratios of change desired in a number of traits are prespecified. Working backwards, the index which would produce these changes is determined.

Restricted indices such as the desired gains index appeal because they avoid the need for estimation of economic values. Against this, much more onus is placed on the breeder and **implied values** are still placed on each trait as a consequence of the nominated desired gains. If the objective is really to maximise profit, a desired gains index will not be as efficient as one directly targeted at economic gain. When it is possible to derive economic value estimates directly, it is therefore generally preferable to use these as the bases for derivation of the selection index.

Guide for Consultants

- Clearly establishing the breeding objective is the first step to effective breeding.
- The process of formally setting the objective can be considered in several logical steps. Using economics to assess the relative importance of the traits to be improved is just one part of this process.
- Using economics to estimate the relative importance of traits in a breeding objective is a specialised application of economics, and special considerations apply. This application should not be confused with other applications, for example in farm management.
- Some principles for estimating trait economic values for inclusion in a breeding objective:
  - Base estimates on the costs and returns of a relevent *commercial* herd or flock.
  - Combine the estimated changes in costs (C) and returns (R) as profit (R-C) or as economic efficiency (C/R), depending on the case. Profit is likely to be the perspective desired for individual breeders, but with care taken that any prevailing constraints on individual or total inputs or outputs are accounted for.
  - Calculate trait economic values assuming that optimal management exists both before and after improvement of the trait.
  - Feed costs especially need to be accounted for, and ideally feed intake should be included as a trait in the breeding objective.
  - Special care should be taken with the units in which economic values are calculated and expressed, and with the way in which the relativity of traits is described.
  - Discounting procedures are best used to adjust estimated economic values for the time delay that occurs before economic benefits are realised from improvement in each trait.
- It is important to be as comprehensive as is practicable in specifying costs and returns, and the traits associated with these, in commercial production. Balanced against this should be the realisation that estimates of trait economic values are all that is possible or desired.
- Complete or partial budgeting methods are usually adequate for estimating trait economic values. When these budgeting methods are inadequate, simulation using bio-economic models may be possible.
- Avoid specifying traits combined as products or ratios whenever possible, as the components of these have uncontrolled values that are implied by their relative contributions to the variation in the products or ratios.
- Restricted indices, such as a desired gains index, avoid some of the difficulty of estimating economic values. However, they utilise implied values which are less efficient than direct estimates if the real objective is economic gain.

References

Atkins KD, McGuirk BJ and Thompson R (1986) Intra-flock genetic improvement programmes in sheep and goats. Proc 3rd World Congress on Genetics Applied to Livestk Prod, Lincoln, IX:605-618

Barlow R (1987) An introduction to breeding objectives for livestock. Proc 6th Conf AAABG, Perth, 9-11 Feb: pp 162-169

Barwick SA and Hammond K (1990) Apportioning emphasis between the seed-stock producer and user in establishing the breeding objective. Proc 8th Conf AAABG, Hamilton and Palmerston North, Feb 5-9: pp 79-84

Brascamp EW, Smith C and Guy DR (1985) Derivation of economic weights from profit equations. Anim Prod, 40:175-179

Dekkers JCM (1991) Estimation of economic values for dairy cattle breeding goals: bias due to sub-optimal management policies. Livestk Prod Sci, 29:131-149

de Vries AG (1989) A method to incorporate competitive position in the breeding goal. Anim Prod, 48: 221-227

Dickerson GE (1970) Efficiency of animal production - molding the biological components. J Anim Sci, 30:849-859

Fowler VR, Bichard M and Pease A (1976) Objectives in pig breeding. Anim Prod, 23: 365-387

Gibson JP (1989) Economic weights and index selection of milk production traits when multiple production quotas apply. Anim Prod, 49: 171-181

Gibson JP and Kennedy BW (1990) The use of constrained selection indexes in breeding for economic merit. Theor Appl Genet, 80: 801-805

Goddard ME (1983) Selection indices for non-linear profit functions. Theor Appl Genet, 64: 339-344

Groen AF (1989) Economic values in cattle breeding I. Influences of production circumstances in situations without output limitations. Livestk Prod Sci, 22:1-16

Groen AF (1989) Economic values in cattle breeding II. Influences of production circumstances in situations with output limitations. Livestk Prod Sci, 22:17-30

Harris DL (1970) Breeding efficiency in livestock production - defining the economic objective. J Anim Sci, 30:860-865

Hazel LN (1943) The genetic basis for constructing selection indexes. Genetics, 28: 476-490

James JW (1978) Index selection for both current and future generation gains. Anim Prod, 26: 111-118

James JW (1982) Economic aspects of developing breeding objectives: general considerations In "Future Developments in the Genetic Improvement of Animals" (Editors: J.S.F. Barker, K. Hammond and A.E. McClintock), p.107 (Academic Press)

McArthur ATG (1987) Weighting breeding objectives - an economic approach. Proc 6th Conf AAABG, Perth, Feb 9-11: pp 179-187

McArthur ATG. and del Bosque Gonzalez AS (1990) Adjustment of annual economic values for time. Proc 8th Conf AAABG, Hamilton and Palmerston North, Feb 5-9: pp 103-109

McClintock AE and Cunningham EP (1974) Selection in dual purpose cattle populations: defining the breeding objective. Anim Prod, 18:237-247

Newman S, Harris DL and Doolittle DP (1985) Economic efficiency of lean tissue production through crossbreeding: Systems modeling with mice 1. Definition of the bioeconomic objective. J Anim Sci, 60:385-394

Ponzoni RW (1988) Accounting for both income and expense in the development of breeding objectives. Proc 7th Conf AAABG, Armidale, Sept 26-29: pp 55-66

Ponzoni RW and Gifford DR (1990) Developing breeding objectives for Australian Cashmere Goats. J Anim Breed Genet, 107:351-370

Ponzoni RW and Newman S (1989) Developing breeding objectives for Australian beef cattle production. Anim Prod, 49:35-47

Rae AL (1988) Including costs in defining objectives for sheep improvement. Proc. 7th Conf AAABG, Armidale, Sept 26-29: pp 67-76

Ronningen K (1971) Tables for estimating the loss in efficiency when selecting according to an index based on a false economic ratio between two traits. Acta Agric Scand, 21:33-49

Simm G, Smith C and Thompson R (1987) The use of product traits such as lean growth rate as selection criteria in animal breeding. Anim Prod, 45:307-316

Smith C (1983) Effects of changes in economic weights on the efficiency of index selection. J Anim Sci, 56:1057-1064

Smith C (1988) Economics of livestock improvement. Proc 7th Conf AAABG, Armidale, Sept 26-29: pp 42-54

Smith C, James JW and Brascamp EW (1986) On the derivation of economic weights in livestock improvement. Anim Prod 43:545-551

Stewart TS, Bache DH, Harris DL, Einstein ME, Lofgren DL and Schinckel AP (1990) A bioeconomic profit function for swine production: application to developing optimal multitrait selection indexes. J Anim Breed Genet, 107:340-350

van Arendonk, JAM and Brascamp, EW (1990) Economic considerations in dairy cattle breeding. Proc 4th World Congress on Genetics Applied to Livestk Prod, Edinburgh, XIV:78-85

Vandepitte, WM and Hazel, LN (1977) The effect of errors in the economic weights on the accuracy of selection indexes. Ann Genet Sel Anim, 9:87-103

Yamada Y, Yokouchi K and Nishida A (1975) Selection index when genetic gains of individual traits are of primary concern. Jpn J Genet, 50:33-41

Chapter 14

Breeding Objectives for Beef Cattle

Stephen Barwick and Willi Fuchs

The Need

Formulating the breeding objective is a crucial step and ideally the first step in any genetic improvement effort. The modern breeding methods introduced to the Australian and other beef industries through the genetic evaluation system **BREEDPLAN** did not evolve in this ideal order, but rather through the initial production of estimates of breeding values (EBVs) for traits currently measured. While the widespread uptake of BREEDPLAN may be evidence for the effectiveness of the approach taken, with farmers first learning about the concept of predicting genetic merit (EBVs or EPDs) of traits, there remains a need for clearer definition of beef breeding objectives, both for individual breeders and for groups representing different types of production or targeting different markets.

This chapter introduces and discusses a breeding objectives computer package recently developed at the Animal Genetics and Breeding Unit for use in beef industries with BREEDPLAN. The package, here termed **B-OBJECT**, is both a potential mechanism for introducing formal objectives to the beef breeding industry and a procedure facilitating multi-trait selection. Some examples are presented. The package also facilitates teaching and research into issues of multi-trait selection in beef cattle.

B-OBJECT: A Beef Breeding Objective and Selection Index Package

A formal breeding objective is a mathematical description of the balance needed between traits for maximum profitability. The B-OBJECT package is designed to help breeders and buyers of seedstock establish their own assessment of this balance, and to provide a means then of selecting for it as simply as possible. The procedure allows the user's own assessment to be developed and utilised, on a customised basis, so that the breeder and buyer have increased decision-making power without loss of control of their economic destinies.

Basis of the procedure

B-OBJECT involves 3 key elements:

- Breeder-supplied production and cost estimates for commercial beef production. These allow assessment of the breeder's required balance between traits, i.e. the breeding objective.
- The measurements, EBVs or EPDs, available to a breeder through BREEDPLAN.

- A description of the inheritance of all traits concerned, including the genetic correlations between all traits and measurements, i.e. the so-called **covariance structure**.

The outcome of the procedure is the best index or combination of measurements to use to breed for the required balance of traits.

Other points to note:
- Without a clearly defined breeding objective it is easy to confuse the traits that are trying to be improved, e.g. female fertility, with those which are just measurements, e.g. scrotal size, providing information about the economically important traits. This hasn't been a problem while BREEDPLAN has dealt only with weights, but with the addition of further EBVs to BREEDPLAN the distinction between traits and measurements is much less clear.
- Use of a covariance structure allows correlated information to be utilised in the same way that BREEDPLAN uses the heritabilities of measures and correlated information to produce EBVs.

Formulating the Breeding Objective

Procedures used in B-OBJECT to formulate the breeding objective are discussed here in relation to the steps usually recommended for establishing the objective.

Identification of the breeding, production and marketing system.

It is first necessary to identify some features of both the herd in which genetic selection is to be practiced and a representative commercial herd in which the improved genes will ultimately be expressed. For the former, or seedstock herd, there is the need to identify the breed and environment involved, as this affects the description of trait heritabilities and genetic correlations which might be utilised, and it is necessary to know the commercial role which improved animals will have in the industry, straight-breeding or a specialised crossing role, since this affects the traits which might be included in the objective. For the commercial herd it is important to know, or to be able to estimate, details of the production system, the target age of turn-off and method of sale, and aspects of the market or markets addressed.

In B-OBJECT, these requirements are met via a questionnaire built into the computer package and completed by the user. The questionnaire caters for answers from quite diverse production systems, since a feature of the procedure is its intent to allow the breeding objective to be customised for individual cases. This desire to provide a facility for customising the objective recognises both that the beef industry involves a vast diversity of breeds, breeding and production systems, environments and markets, and that breeders need a strong sense of ownership of any objective if they are to pursue it.

In practice, differences between the indexes for individual breeders may often be small. The largest differences are expected when there are differences in the traits in individual objectives as well as differences in the importance of the traits. One difficulty is that common covariance structures sometimes have to be used for deriving indexes in situations where differing ones might preferably be used. This works to restrict the differences seen between indexes. It is necessary, however, because of the scarcity of reliable sets of genetic parameter estimates. Ultimately, it may be possible to use completely breed-specific covariance structures. The capacity for customising the selection index will consequently increase as more and better estimates of the necessary parameters become available.

A first categorisation of breeding situations for which covariance structures might be available is shown in Table 14.1. A further development would be to distinguish between markets on the basis of the economic importance of marbling. Example objectives and indexes derived for the categories of cases shown in Table 14.1 might provide useful bases from which modifications could be made for individual breeder cases.

Table 14.1 A First Categorisation of Breeding Situations

Environment	Genotype	Breeding Function	Commercial Age of Turn-off
Temperate	Bos taurus	Straight-breeding	10m or less ('Vealers', V) More than 10 to less than 17m ('Yearling', Y) 17m or more ('Ox', O)
		Terminal-Sire[1]	V,Y and O cases
		Maternal[1]	V,Y and O cases
Tropical/Sub-Tropical	Bos indicus	Straight-breeding	V, Y and O cases
		Terminal-Sire	V, Y and O cases
		Maternal	V, Y and O cases
	Zebu-derived	Straight-breeding	V, Y and O cases
		Terminal-Sire	V, Y and O cases
		Maternal	V, Y and O cases

[1] Specialised roles in crossbreeding.

Identification of sources of returns and costs in commercial herds.

Returns are influenced by the number of animals sold, by the weight of each and by the price per kilogram of each, for each class of sale animal - normal sale animals, including any receiving a price penalty for not meeting particular specifications, culled and aged cows and culled and aged bulls. Costs include feed costs, husbandry costs and marketing costs for all relevant classes of stock. Feed costs are a function of the amount eaten and the cost per kg of feed. Fixed costs, those which don't vary with the level of production, may also need specifying if something other than change in unconstrained profit is used as the basis for assessment of trait economic values.

An issue in specifying the sources of returns and costs is the level of detail which should be attempted. For returns, this can be guided by the way marketing is visualised as occurring over the time horizon of the genetic improvement effort. In addition to returns per animal or per carcase, it might be foreseeable, for example, that returns might be separately attributable to the amount of lean meat in the carcase, or to differences in the amounts of individual carcase cuts.

Determination of the biological traits influencing returns and costs.

Determining the biological traits associated with each source of returns and costs in commercial herds is a crucial step in setting the breeding objective. Here, the concern is for a clear association with either or both of returns and costs and for all returns and costs to be accounted for by the traits determined. That is, the specification of the objective needs to be comprehensive. Ideally, this should extend to including traits that have importance through their effect on performance in the current herd as well as those that are expressed in progeny.

The exact specification of the traits required may be interpreted differently by different individuals, usually without serious consequence. As with the sources of returns and costs, most differences arise from the level of detail attempted in the specification. Common reasons for differences are the specification of some traits separately for different sexes or age classes of animals, or the expression of traits in composite, rather than component form. Again, the basis on which marketing is expected to be carried out over the time horizon of the genetic improvement effort can be used as a guide to the level of detail required in specifying traits.

A further point is that the objective may be little affected by the exclusion of traits which account for only a very small proportion of profit, at the genetic level, e.g. when the product of the trait economic value and heritability is trivially small. In these instances, a failure to account for some aspect of either returns or costs may be of virtually no consequence. Such omissions need to be carefully evaluated, however, before being countenanced.

Issues that are not completely resolved include whether calving rate, or a more composite trait such as weaning rate, or days to calving should be the female reproductive trait specified in the objective, and how current herd aspects of traits like mating potential of bulls should be combined with breeding values to yield a more comprehensive definition of the objective. Whether or not a cow calves or weans a calf is easily understood to be the important economic trait for female reproduction. However, it is a very coarse measure, there being only two possible calving outcomes for any breeding opportunity. Days to calving is a more continuously-distributed variable and it may be the variable which underlies differences in calving rate, the argument being that all cows would calve if they had sufficient time. Fortunately, differences do not appear to be large between indexes formulated for objectives with female reproduction specified as either calving rate or days to calving.

A further issue is the manner in which feed costs are accounted for in the specification of the objective. The correct procedure is to include feed intake as an objective trait. However, to obtain the necessary genetic parameter estimates for feed intake including genetic correlations with other traits is itself a massive undertaking. Consequently, an alternative which is sometimes used is to estimate feed requirements from liveweight and to consider increased feed requirement as a cost associated with increase in weight. This in effect assumes a high phenotypic correlation between feed intake and weight and zero phenotypic correlations of feed intake with other available measures.

Traits considered in the objective in Version 1.20 of B-OBJECT are as follows:
- sale weight (direct and maternal)
- dressing percentage
- saleable meat percentage
- fat depth
- cow weaning rate
- bull fertility
- cow survival rate
- cow weight

Sale is assumed to occur when animals are at the targeted sale age. Liveweight is utilised as the basis for estimating feed requirements. The package allows flexibility in trait definition as regards the sale weight and carcase traits, differing parameters being allowed, for example, for sale weight depending on the age at sale.

Derivation of the economic value of each trait

The breeding objective can usually be represented as the sum of the true breeding values (BV) of a number of traits, each multiplied by a factor indicating its importance, e.g.

$$v_1(BV_1) + v_2(BV_2) + v_3(BV_3)$$

The factors indicating relative importance, i.e. the vs above, are usually economic values. Some of the issues in estimating these values are discussed in Chapter 13. In B-OBJECT, economic values are estimated as the change in profit associated with a unit of change in each trait, for a relevant commercial beef enterprise. Values for each trait are assessed assuming the other traits of the objective are held constant. Options are available for computing values assuming either that extra feed is able to be purchased when needed or that the total available feed is fixed. Under the latter assumption, extra requirement is met by reducing stock numbers. Assessment of economic values are based on the breeder's own estimates of production and cost characteristics for commercial beef production. The seedstock breeder makes these estimates for a typical commercial herd in which his bulls will be used.

When additional feed can be purchased, the economic value for sale weight is estimated as the difference between the value of a kg of finished sale animal and the cost of the extra feed needed to produce the extra kg, the estimate then being discounted to present value. The value for sale weight maternal is similar to that for sale weight direct except that the maternal trait is discounted more on account of its later expression in the herd.

The economic value for dressing percentage is the value of the increased weight of meat resulting from a one percent increase in the amount of the live animal that is represented as carcase. The economic value for saleable meat percentage is the value of the increased weight of meat resulting from a one percent increase in the amount of the carcase that is represented as meat.

The value of an extra millimetre of fat depth, when meat yield is constant, is based on the reduction in incidence of sale animals which fail to meet minimum fat specifications for the targeted market, and from the price penalty incurred for failing to meet this minimum specification.

The economic value for weaning rate derives from the additional profit accruing from an extra one percent of calves after accounting for additional feed costs and other costs, and from additional effects on profit that arise from changes in the age structure of the herd. The age structure is affected because fewer cows are culled because there are less dry cows.

Bull fertility is valued by the reduction in the average bull cost per cow that occurs when bulls are able to satisfactorally service an additional cow.

The value of improving cow survival rate derives directly from the altered age structure that results from fewer cow deaths. This affects the numbers of cull and cast-for-age cows available for sale, and the number of replacements needed from the young females that would otherwise be normal sale animals. The differing costs of feeding and running the changed herd structure are also taken into account.

Increasing cow weight increases revenue through the sale of heavier cull and cast-for-age cows. There is an additional feed cost, however, in running heavier cows. The economic value of cow weight is assessed as the difference between this additional revenue and additional cost.

Economic values for all traits in B-OBJECT are discounted to account for the differing time delays before expression of the improvement in the herd, and are expressed on a per cow basis.

Deriving the selection index

Available criteria for the index. The measurements or BREEDPLAN EBVs or EPDs that are available to the breeder are the sources of information, or potential selection criteria, to be combined in the Index. The breeder nominates the EBVs which are available, and this decides what can be included in the index. EBVs that B-OBJECT presently caters for as criteria are

Birth Weight
200-Day Weight (Milk)
200-Day Weight (Growth)
400-Day Weight
550-Day Weight
600-Day Weight
700-Day Weight
900-Day Weight

Days to Calving
Scrotal Size
Fat Depth
Eye Muscle Area

With only a couple of exceptions, e.g. breeders either have 400 and 600-Day EBVs or 550, 700 and 900-Day EBVs from BREEDPLAN, any combination of EBVs can be nominated as available for the purpose of index construction.

Assumed genetic parameters. A matrix of genetic parameters, including correlations between all of the traits in the objective and all of the selection criteria, underlies the derivation of the index for any given objective. By taking account of the genetic variability in each measure and trait, and of the correlations among these, the best weighting to apply to each EBV is arrived at, this combination being the best single estimate of the breeding objective. Where, for example, four BREEDPLAN EBVs are the selection criteria, the resulting index takes the form

$$b_1(EBV_1) + b_2(EBV_2) + b_3(EBV_3) + b_4(EBV_4),$$

the **b**s being the index weightings.

Deriving the index weightings. It was shown in Chapter 13 that when observed measurements are the information available for construction of a selection index, the equations to be solved to obtain the most appropriate weightings for each measurement can be written as

$$\mathbf{Pb = Gv}$$

where
- **P** is the phenotypic variance-covariance matrix among measurements,
- **G** is the genetic variance-covariance matrix between the measurements and the traits in the objective,
- **b** is a vector of the weightings to be solved for,
- **v** is a vector of economic values of traits,

and the index weightings are given by

$$\mathbf{b = P^{-1}Gv}$$

In B-OBJECT, EBVs are the criteria to be combined, rather than observed measurements. Under these circumstances, the analogous solutions are obtained as

$$\mathbf{b = G_{11}^{-1}G_{12}v}$$

where
- $\mathbf{G_{11}}$ is the genetic variance-covariance matrix among the criteria in the index,

and
- $\mathbf{G_{12}}$ is the genetic variance-covariance matrix between the criteria in the index and the traits in the objective.

B-OBJECT Applications

Who will use B-OBJECT?

Potential users are:

- Bull breeders:
  - As a selection aid, utilising the most appropriate single genetic ranking and using this in association with other culling criteria.
  - As a marketing aid, providing separate rankings for different individual clients or types of clients.
- Bull buyers, individual buyers or buyers grouped according to their type of beef enterprise:
  - As a buying aid, utilising the most appropriate ranking of available bulls for the required production system and targeted market.
- Industry consultants:
  - To help match bull-buying clients with available sale bulls.
  - As an aid for both bull-breeding and bull-buying clients.
- Breed societies and other users of BREEDPLAN International:
  - To promote bulls, and breeding practices, for specialised functions or for versatility over several beef production systems and markets.

Ultimately, the whole industry will benefit from the enhanced capacity to produce the beef required for specific markets.

Further reasons for using B-OBJECT

- Breeders can use the combination of EBVs which gives them most genetic progress for the least recording effort and processing cost.

  Questions about how important individual EBVs are, whether they are ones that are already in BREEDPLAN, or ones that might be, are of obvious importance. They can only be answered for a defined breeding objective. Once this balance between traits desired by the breeder is known, the B-OBJECT approach makes it possible to estimate the usefulness of any existing or proposed EBV.

- Breeding can be targeted more precisely at markets, and for the flexibility to service several markets.

  The capacity to target markets, and to be able to access the genetic material to do this, is being increasingly rated by commercial breeders as their area of greatest information need. This is a central aim of B-OBJECT. With B-OBJECT, single genetic rankings for different purposes, e.g. production of different types of commercial product can easily be produced.

- BREEDPLAN EBVs or EPDs can be used more efficiently.

  Breeding involves more than one trait. B-OBJECT facilitates multi-trait selection by providing the mechanism to best utilise available BREEDPLAN EBVs.

- BREEDPLAN can increasingly be directed at the needs of the commercial sector.

    The B-OBJECT approach allows separate rankings for different types of commercial producers. These rankings could be appropriate for sale catalogues. Rankings can also be provided for individual commercial breeders.
- Management of BREEDPLAN information is made easier.

    As more EBVs are added to BREEDPLAN, it becomes more likely that they will rank animals differently. No longer will EBVs themselves be highly correlated as has been the case with growth EBVs. B-OBJECT provides a simple way to combine EBVs for specific purposes.
- Developers of BREEDPLAN will be assisted

    When used with BREEDPLAN, B-OBJECT provides a valuable model of the genetic improvement process. It has potential as a research and educational tool. It will aid the early detection of any weaknesses in multi-trait prediction, and help encourage efforts from both breeders and BREEDPLAN developers in the most profitable direction.

B-OBJECT Examples

Example calculations of economic values for traits in the objective are shown in Table 14.2 for Vealer, Yearling and Jap.-Ox cases of commercial beef production. There can be considerable variation around these values for individual breeder cases. Recall that it is the **relativity** of the economic values that affects the breeding direction and that this relativity is best appreciated for a standard amount of the variability usually, a phenotypic standard deviation of each trait.

Table 14.2 Economic Value Estimates for Traits of the Breeding Objective for Example Cases of Commercial Beef Production, calculated on a per cow basis

Trait	Unit	Vealer	Yearling	Jap.Ox
Sale Weight (Direct)	kg	0.85	0.67	0.50
Sale Weight (Maternal)	kg	0.52	0.40	0.30
Dressing %	%	4.26	5.55	6.58
Saleable Meat %	%	3.37	4.38	5.09
Fat Depth (rump)	mm	1.68	0.79	0.98
Cow Weaning Rate	%	1.58	2.21	2.56
Bull Fertility	mate	0.12	0.11	0.10
Cow Survival Rate	%	2.43	2.73	3.20
Cow Weight	kg	-0.13	-0.12	-0.12

Index weightings for each of the same example cases are shown in Table 14.3. These are based on estimates of genetic parameters for the Australian industry. The apparent importance given to each EBV, by these weightings, is also best appreciated by taking account of the variability inherent in each EBV criterion. This resulting assessment is illustrated in percentage terms in Table 14.3 by the figures described as **index composition**.

The indices of Table 14.3 were each applied to BREEDPLAN EBVs for a real herd. The individual EBVs and overall rankings for a Yearling beef production objective are shown in Table 14.4 for the top 15 of 81 young bulls available. The comparative rankings of the same young bulls under each of the example objectives are given in Table 14.5. Notice that a very substantial amount of re-ranking occurs for different commercial uses of bulls.

Table 14.3 Index Weightings for EBVs and the Approximate % Composition of Indexes for Example Cases of Commercial Beef Production.

EBV	Vealer Index Weighting[2]	Vealer Index Comp.	Yearling Index Weighting[2]	Yearling Index Comp.	Jap. Ox Index Weighting[2]	Jap. Ox Index Comp.
Birth Weight	0.016	0%	-0.103	1%	-0.358	3%
200-day Milk	0.308	12%	0.094	3%	-0.128	4%
200-day Growth	0.501	23%	0.187	8%	0.095	3%
400-day Weight	0.114	7%	0.276	16%	0.175	9%
600-day Weight	-0.054	5%	0.031	2%	0.118	8%
Days to Calving	-0.995	27%	-1.363	33%	-1.539	32%
Scrotal Size	0.499	3%	1.079	6%	2.080	10%
Fat Depth	-2.253	12%	-3.770	18%	-4.499	18%
Eye Muscle Area	0.764	10%	1.193	13%	1.449	14%
Accuracy[1]		0.22		0.24		0.26
s_I[1]		3.99		5.58		6.59

[1] Assuming an accuracy for individual EBVs equivalent to the square root of the heritability

[2] Direction of the change given by the sign of each weighting.

Table 14.4 GROUP BREEDPLAN EBVs and $INDEX-EBVs for Young Bulls for an Example Yearling Beef Production Objective

ID	BW	200M	200D	400D	600D	DC	SS	FD	EMA	$ INDEX
517	+1.6	+5	+6	+21	+12	-1	+0.4	-0.3	+2.7	+14
541	+2.6	+5	+9	+26	+34	-1	+0.8	+0.4	+1.8	+13
490	+1.6	+2	+5	+23	+23	-2	+0.3	+0.2	+0.7	+11
644	+2.1	+3	+6	+18	+26	-2	+0.5	-0.2	+0.1	+11
606	+3.4	+4	+8	+28	+34	0	-0.3	+0.0	+0.5	+11
521	+0.1	+2	+3	+24	+28	0	+0.5	+0.3	+2.3	+10
512	+2.6	+5	+6	+16	+16	0	-0.1	-0.3	+2.4	+10
645	+0.5	+5	+5	+17	+18	-1	+0.4	-0.1	+1.1	+10
548	+1.5	+2	+6	+18	+19	-1	+0.6	-0.1	+0.1	+9
467	+1.4	+2	+6	+15	+17	-1	+0.3	+0.0	+1.0	+9
528	+1.9	+6	+6	+11	+15	-1	+0.0	-0.2	+1.1	+8
591	+2.7	+6	+7	+16	+22	0	-0.4	+0.6	+2.9	+7
511	+1.7	+1	+5	+24	+24	-1	+0.0	+0.4	-0.7	+7
621	+0.9	+6	+4	+15	+3	-1	+0.6	+0.1	+0.1	+7
642	+2.3	+3	+7	+23	+20	0	+0.5	+0.5	-0.5	+6

Table 14.5 Comparative Rankings of Young Bulls for Different Example Cases of Commercial Beef Production [1,2]

Vealer v Yearling		Vealer v Jap. Ox		Yearling v Jap. Ox	
517 (+10)	517 (+14)	517	517 (+12)	517	517
541 (+9)	541 (+13)	541	541 (+12)	541	541
512 (+8)	490 (+11)	512	644 (+11)	490	644
528 (+8)	644 (+11)	528	521 (+11)	644	521
644 (+7)	606 (+11)	644	490 (+10)	606	490
645 (+7)	521 (+10)	645	645 (+9)	521	645
606 (+7)	512 (+10)	606	548 (+9)	512	548
490 (+7)	645 (+10)	490	512 (+8)	645	512
591 (+7)	548 (+9)	591	606 (+8)	548	606
621 (+7)	467 (+9)	621	467 (+8)	467	467

[1] IDs underlined are not present in the list with which they are compared. Actual Index values are shown in parentheses.

[2] Target sale ages: Vealer - 9m; Yearling - 14m; Jap. Ox - 24m.

Guide for Consultants

Whether selecting animals on visual traits only or selecting them using a combination of EBVs and visual selection, the breeder makes intuitive decisions about:

- The relative value of a range of traits e.g. growth rate, scrotal size, structural soundness, eye pigmentation etc. These economic values depend on the market supplied.
- Whether traits are genetically associated, e.g. whether increasing growth rate will have an effect on dystocia, or greater muscling an effect on fertility etc.

B-OBJECT is a computer package which can help breeders make these important decisions more objectively and more accurately, i.e. it is a further decision-aid for breeding. It can be used by both bull breeders and bull buyers to:

- Suggest the relative weighting that should be put on the measures or EBVs that are available for use in selection, and
- Rank animals for their overall breeding value for a given type of commercial production system and market

It can also be used to predict ahead what genetic changes will occur in a herd if a particular set of weightings are applied to selection of the herd, to compare alternative selection strategies, and for testing how sensitive any strategy is to a change in market requirements or prices.

Beef industries have a history of chasing and promoting some kind of maximum, e.g. maximum weight, maximum height or maximum muscle. Yet many breeders consider we should be breeding for an **optimum** rather than for a maximum. Saying we should breed for an optimum is another way of saying that we should be selecting in a balanced way. But what is the balance that is needed between traits for maximum profitability, i.e. the breeding objective? B-OBJECT is designed to help breeders establish their assessment of this balance, and to provide a means then of selecting for it as simply as possible. Note that the package is designed to allow the breeder's own assessment to be developed and utilised, on a customised basis, so that breeders retain control of their own economic destiny.

B-OBJECT also has potential as a teaching and learning package for breeders, extension staff and students.

Steps in Using B-OBJECT

Interactively with a consultant:

Step 1: Breeder completes questionnaire on-screen. This includes nominating which EBVs or other measures are available and production and cost estimates for a target commercial beef enterprise i.e. production system and market.

Step 2: Economic values are calculated and the breeding objective formulated.

Step 3: A matrix of assumed genetic correlations between all traits in the objective and all selection criteria is utilised and the most appropriate index of the available selection criteria is derived.

Step 4: The program is told where the selection criteria details, e.g. the breeder's EBV or EPD file, can be found for the animals of interest. The index is then calculated for each animal.

Step 5: The results are available for display on screen or for printing. These can be sorted and the output restricted so that, for example, long-gone animals do not appear. Component EBVs of the index are also displayed.

Numerous other features are also provided to safeguard against inconsistent input, provide what if capability, and to add to the usefulness of results. A summary of features is given in Figure 14.1.

References

McArthur ATG, del Bosque Gonzalez AS (1990) Adjustment of annual economic values for time. Proc 8th Conf, AAABG, Hamilton and Palmerston North, 103-109

Ponzoni RW, Newman S (1989) Developing breeding objectives for Australian beef cattle production. Animal Production 49: 35-47

Schneeberger M, Barwick SA, Crow GH, Hammond K (1992) Economic indices using breeding values predicted by BLUP. Journal of Animal Breeding and Genetics (In press)

Upton WH, McArthur ATG, Farquharson, RJ (1988) Economic values applied to breeding objectives: a decentralised approach for BREEDPLAN. Proc 7th Conf, AAABG, Armidale, 95-104

Client Capacity
- Provides for storing up to 15 questionnaires (breeders) at one time

Interface with BREEDPLAN
- Any combination of available EBVs can be nominated to be used in the index

Interactive Inputting
- Questionnaire available on-screen, and can be completed with breeders interactively
- Help function available
- Input range checks and consistency checks
- Capacity for flagging inputs
- Graphic display of timetable of input variables

Displays of Intermediate Results
- Economic values for traits in the breeding objective
- Index weightings
- Index composition
- Expected genetic gains in each trait in the objective and in each EBV criterion, both in units of measurement and in $units.
- Other help screens

What if you Don't Agree
- All operations are quickly and easily repeated
- Capacity to by-pass the computation of economic values by inserting other values
- Capacity to modify the composition of the Index directly

Using the Finalised Index
- The breeder's EBV file is easily read and the Index value computed for all animals
- Numerous file formats can be handled
- Comparison of selected and total animals for any nominated % selected

Outputting Results
- 3 sort options available
- Capacity to limit output by the sort option variables and by age/sex category
- Graphic display of genetic trend
- Printed report facility available
- Breeder-supplied production and cost estimates can be printed seperately

System Requirements
- IBM-compatible micro-computer with 640K RAM
- Hard disk drive with approx. 2 megabytes of free space (desirable)
- Colour or Monochrome EGA (Enhanced Graphics Monitor) monitor and card (desirable)

Fig. 14.1   Some features of the B-OBJECT package.

Chapter 15

Breeding Objectives in Wool Sheep

Mick Carrick

Direction of Genetic Change is Important

Genetic change can be a very potent tool for increasing net income from livestock enterprises. However it can be, and often is quite ineffective or even deleterious because all of the necessary steps have not been carried out in detail or because important quality traits have been ignored. There is no point in achieving rapid genetic change if its direction is not that most likely to improve or ideally to maximise the rate of increase of income from breeding. Rapid genetic change may also be deleterious if the breeding objective has been oversimplified.

The formal approach to setting breeding objectives is an important tool in the craft of designing breeding programs to maximise profit in any breeding system.

It is, however, not the sole ingredient in the recipe for good design of breeding programs and we should not assert, as has been done in some notorious extension program, that this approach is the whole answer. In practice, we must integrate the formal approach with less mathematically tangible elements such as maintenance of satisfactory functionality of anatomical structure and, in certain markets, wool quality traits other than fineness.

Which Approach to Use

There are two fundamental ways to think about breeding objectives: the formal approach to be covered here in detail and the **desired gains** approach in which the formal approach can play a role. Here we effectively pre-judge what the desired outcome of the program is to be and design the selection methods to meet this objective. This has a lot of appeal in the breeding community because the objectives stated are usually ones that the breeder is already comfortable with. In some instances, which will be covered later, it is the only way to cope with objectives.

The desired gains approach can be evaluated in parallel with the formal approach to let the breeder know if his objectives deviate significantly from one which maximises increase in profit or in other words whether he can increase his profit by deviating from the approach which seemed intuitively correct.

The main criteria for good selection objectives are that they be both clear and realisable. Each of the two methods of setting objectives can result in clear objectives but the desired gains approach can sometimes produce objectives which are not genetically realisable. There is little point in setting an unreachable goal.

The Formal Approach Aims to Maximise Profit Increase

The formal approach lists the traits the breeder wants to change and analyses the relative improvement in profit brought about by changing each of them. We use a whole-farm gross margins approach to this analysis which takes into account both increases in product values as well as costs associated with changes in each trait. If one simply wants to ensure that the correct relative economic values for traits are included, an appropriate partial budget will do, however, to predict the future economic value of the breeding program based on an index then a full budget is necessary.

An example:

When evaluating the increased profit from getting an extra lamb from a ewe, we should take into account the value of extra surplus animals as well as the costs of extra feed for the ewe during pregnancy and lactation and of lost wool income during these phases. We can also account for the increased fibre diameter and lower clean fleece weight of the small proportion of extra twins which accompanies increased fertility.

The formal approach makes use of information on flock structure, costs and incomes in the enterprise and flock breeding history to decide on the relative economic value of the traits which affect income and costs. The appended set of questions for breeders gives the necessary range of input data for calculating the right set of relative economic values.

The most crucial elements of the objective for wool sheep are those which affect profit as follows:

- to increase clean fleece weight
- to decrease average fibre diameter
- to increase lamb weaning percentage and
- to increase sale body weights.
- to increase disease resistance

Of these, fleece value components are currently by far the most valuable, but it is possible to incorporate many other traits if these are important in particular environments or market segments. For example, if the average fibre diameter of the flock or flocks is 17 or 18 microns, we may find that improvements in wool style and colour have a significant impact on income through premiums. If average fibre diameter is 21 or more microns, there is little if any premium for style. Location can also affect the relative values of elements in the objective. In different markets there appear to be somewhat different sets of price differentials for fineness and this can have an impact on the prices obtained for lower micron wool.

The outcome of the formal approach is a list of traits in the objective, each of which has a dollar value for a unit change independent of the others. For example we might be able to state that in a particular breeding program, environment and market an extra kilogram of clean fleece weight has a net value of $6.50 per ewe/year, reducing fibre diameter by one micron has a net value of $6.95 per ewe/year and so on for the value of extra lambs and body weight. Our objective in the breeding program then becomes a matter of increasing the combined total value of these elements through breeding at the maximum rate.

Integrating economic value and genetics

Having arrived at a functional selection objective for a particular purpose, we can then begin to design a breeding program which will achieve it. The first step in this is to derive the appropriate selection index and/or independent culling levels for the recorded traits. Here we will see the power of the index approach to achieving the objective. To combine economics and genetics, we also use the variances and covariances (associations) among the traits and the measurements being recorded.

Heritability - How easy to move traits

What we have at this point, is a list of traits to improve and an estimate of their economic values in the system independent of each other. However we know from genetic research that not all the traits are equally heritable, traits differ in the degree to which observed superiority is passed to offspring. For example, clean fleece weight has a heritability in most populations of approximately 30% (or 0.30), which means that of all the variation seen in clean fleece weights about 30% is useable genetic variation. It also means that if we take a group of animals which are say 2 kg clean fleece weight superior to average of the flock then we can conclude that they are genetically about 0.6 kg superior to the average. As another example, we know that in many populations the heritability, the degree to which observed superiority is passed to offspring, of fibre diameter variation is approximately 50% (or 0.50). This means that it is nearly twice as easy to shift fibre diameter genetically than clean fleece weight. So the right combination of these traits to maximise the increase in fleece profit is not only weighted by the dollar value of traits but also by the ease with which they can be changed.

Genetic correlation - How one trait tends to move another

There is a second genetic consideration in getting the right answer and this involves what are called genetic correlations. A genetic correlation is a measure of the tendency, by no means absolute, for an animal which is genetically superior for one trait to be genetically superior or inferior for another. As an example we can take the same two traits as above. In general, many of the high clean fleece weight hoggets in a mob will also be above average fibre diameters. This means that if we selected for increased clean fleece weight only, we have a tendency to increase fibre diameter to some extent as a result of the genetic correlation. The strength of this correlation is only around 0.25 in medium wool populations, so there are many individual exceptions to the rule that higher clean fleece weights have higher average fibre diameters. These sheep with genetically high fleece weight combined with genetically below average fibre diameters are our real flock improvers if we are trying to breed for increased fleece value.

The use of genetic correlations also comes into play when, for some practical reason, we cannot directly measure one or more of the traits we want to improve. For example we can make use of the fact that the genetic correlation between greasy and clean fleece weight is quite high (about 0.75) and measure only say greasy fleece weight and fibre diameter when we really want to improve clean fleece weight and fibre diameter. As a further example, much of the benefit of increased body weight comes about through a favourable genetic correlation between ewe body weight and lamb weaning percentage.

Variability - How much room to move traits

The final element in the set of equations to solve is the amount of variability of the various traits because this tells us how much selection pressure we can place on each trait. To use the same examples as above we find that in the average flock 95% of the ewes will have a clean fleece weight between about 1 kg above or below the average of the flock, that 95% of ewes have a greasy fleece weight between 1.3 kg above or below the average, and 95% of ewes have a fibre diameter between 3.5 microns above or below the average. The most common measure of variability is called the standard deviation and to put that into perspective we expect 95% of animals to measure between 2 standard deviations above or below the average.

What we are really interested in for breeding success is the total amount of genetic variation available for selection and how well or accurately we can estimate the genetic merit of an individual from its measured performance. Note that fibre diameter is very variable in flocks and the high heritability tells us that we can make use of this variation effectively in a breeding program. We could have combined the phenotypic variabilities with heritabilities and then dealt directly with genetic variances and covariances.

Finalising the selection index

To put it all together: we know the value of changes in the quantity or quality of products in the particular system, how easily they can be changed genetically, how much a unit of change in one trait will affect change in another and finally how much variability we have to work with in the population. Figure 15.1 shows how the elements go together in a simplified example of aiming to improve clean fleece weight and fibre diameter and measuring greasy fleece weight and fibre diameter.

In this simplified example, the index is a weighted combination of an individual's deviation from the average greasy fleece weight and average micron which will maximise the rate at which the total value of the fleece increases. The central idea is to find the most profitable compromise set of attributes - a somewhat lower greasy fleece weight combined with a superior micron can have a higher index value than one with high greasy fleece weight with a broader micron.

An index can also be thought of as a single trait with its own heritability, and in this instance it is the trait with the highest dollar value for breeding.

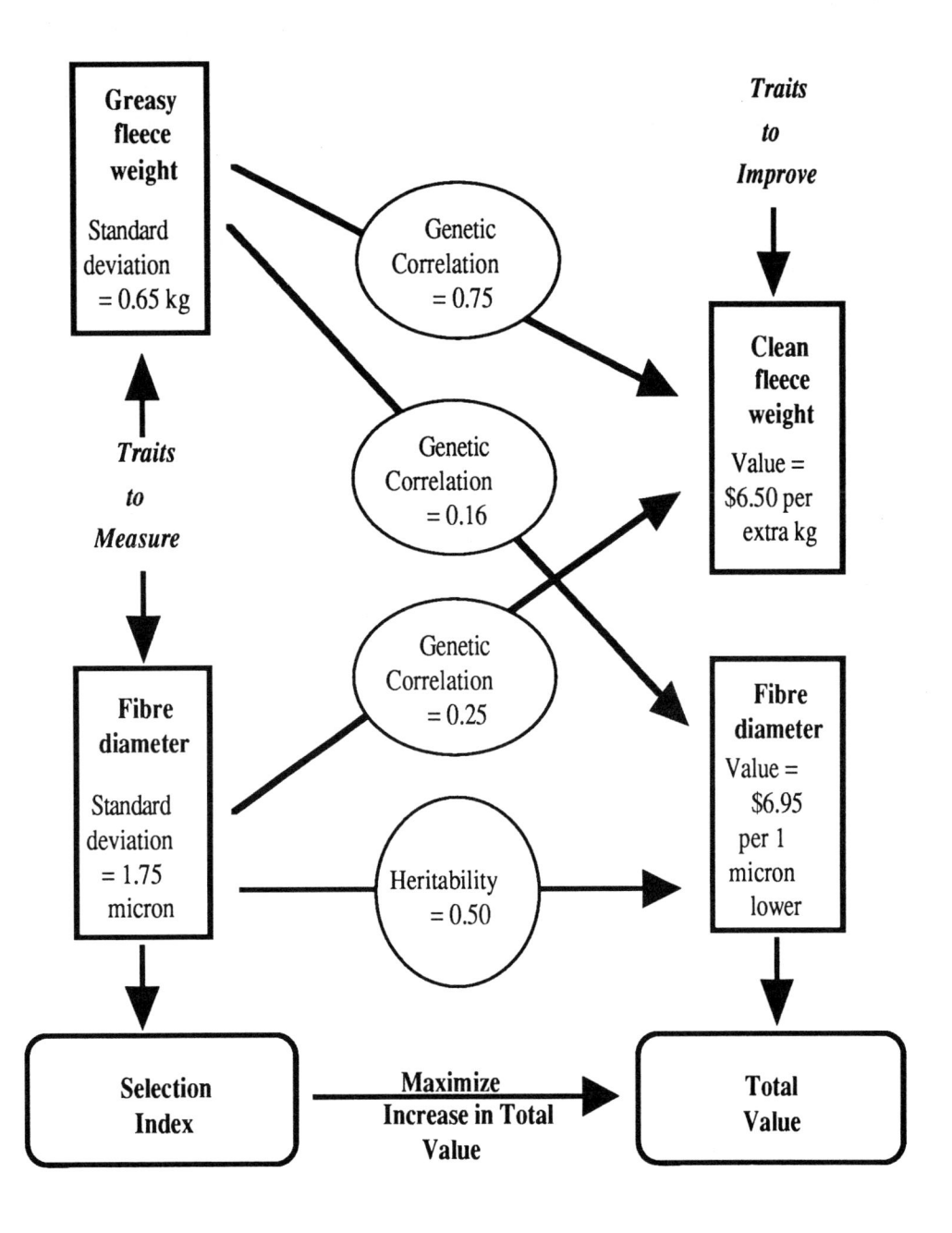

Fig. 15.1 A reduced example of the elements used in deriving a selection index:

The Data needed from the breeder

In order to calculate the correct dollar values for the various traits which the objective is to improve we need to have information from the breeder on a range of statistics from his flock. It is best to gather these figures as longer-term averages, such as a five-year average. Particularly during periods of turbulent market changes, a short-term set of figures can be most misleading, especially when we bear in mind that the results of the calculations will be used to produce longer term changes in the flock. Begin by taking a breeding history of the flock to provide an idea of the starting point.

The flock structure, how many males and females of what ages are kept in the flock, is important in allowing estimation of the number of times which a trait is expressed before an animal is disposed of. It is also used to determine what selection differentials can be achieved and at what generation interval.

Current levels of production give an estimate of the combination of genetic and environmental effects in the current situation and should include reproduction rates. The value of production for each part of the enterprise which produces income is important as well as asking for an estimate from the breeder of his or her expectation about the value of changes in value with changes in productivity.

It is also important to find out what a breeder's target levels of production are. In some cases these may be very sensible and achievable. But in cases where these differ markedly from the optimum economic objective or are not genetically possible in the time frame, then explanations about these differences are necessary.

Sample forms for gathering these data from breeders follow.

Breeding Objective Data

Name _____

Address _____

_____

_____ Postcode _____

Trading Name _____

Background to your Breeding Program

Breeding program commenced in _____

Bloodline commenced on _____

Introductions:

Bloodline (source)	Ewes	Rams
Used:		
Occasionally		
Every 2-5 years		
Every year		
Last time		
Extent of use:		
Not retained		
Special family		
Used extensively		

Mating Strategy

Natural Joining	_____	% ewes involved
Artificial Insemination	_____	% ewes involved
Fresh semen	_____	% ewes involved
Frozen semen	_____	% ewes involved
Embryo Transfer	_____	% ewes involved

Rams Supplied :   Number of clients   _____

Total number of rams supplied   _____

Please attach an extra sheet for history if this is inadequate

Breeding Objectives and Relative Economic Values

The purpose of this questionaire is to provide data on flock structures, breeding histories and costs and prices so that we can generate the most secure and profitable breeding objective and the appropriate selection indexes to achieve it most efficiently.

These questions help us know the following important statistics:

1. The current flock structure

2. The current levels of production

3. The value of production

4. The target levels of production

and are outlined in the following tables. Please use actual data where possible and indicate estimates with an (E).

1. The Current Flock Structure:

This information relates to the number of animals in the flock, their ages and the length of time they spend in the flock. All of this helps to determine both the economic values and the expected rates of genetic change under selection.

Please try to do this for the average or most representative commercial flock in the system, and let me know how many flocks exist at each level and their average sizes.

Class	Number	%Kept	Deaths	Number Sold	Age at Sale	Number of times shorn
Weaner Ewes Wethers						
Hogget Ewes Wethers						
Adult Ewes Wethers						

Definitions:

Weaner A young sheep, male or female, up to the age of first shearing (9 to 10 months)

Hogget A sheep, male or female, at first test shearing (10 to 15) months

Adult A sheep, male or female, including maidens that is old enough to be used or selected into the breeding flock (includes maidens)

Times Shorn Is the number of times the individual sheep is shorn as a member of each class (for example once as weaners, once as hoggets and four times as adults)

2. The Current Levels of Production:

A. Wool Production

The levels of production should represent the weighted average of the last five years or more if available. This information is available on the account sales returned to the grower by the woolbroker. Style and length are optional.

Class	Greasy Fleece Weight kg	Yield %	Micron	$ per Head Sold	Style	Length A/B/C
Weaner Ewes Wethers						
Hogget Ewes Wethers						
Adult Ewes Wethers						

Definitions

Greasy Fleece Weight	All wools for each class of sheep (all fleece lines including second and cast lines)
Yield	All wools for each class of sheep (all fleece lines including second and cast lines)
Micron	All wools for each class of sheep (all fleece lines including second and cast lines)
$ per Head sold	Use the average clean price times the average clean fleece weight
Style	Relates to AWC styles i.e. spinners, best topmakers, etc
Length	Staple length expressed as A, B or C length wools

B. **Reproduction Rate**

This information should be the average reproductive performance of all age groups of lambing ewes over the last five years or more.

Class of Sheep	No. of Joinings	% Lambs Marked per Ewes Joined
Adult Ewes		

Definition:

No. of Joinings   The number of times a ewe is joined whilst in the flock

3. The Value of Production:

A. Sale of Surplus Sheep

The values given should reflect the average prices received over five years or more to help even out short term fluctuations.

Class	$ value per Head Lasy Five Years	$ value per Head Next Five Years	Wool Growth in Months

Weaner
  Ewes
  Wethers			

Hogget
  Ewes
  Wethers			

Adult
  Ewes
  Wethers			

Definition:

Woolgrowth in Months   The Number of months growth of wool at the time of sale e.g.. 6 months, 12 months etc

B. Value of Added Production

The information provided here should be estimates of the value of extra production in one trait without any change in another. For example, one extra kg of clean wool at the same micron.

Change	$Value - Last Five Years	$Value - Next Five Years
One extra kg clean fleece per ewe		
One micron lower		
One extra lamb weaned per ewe		

Definition:

One extra kg of wool How much extra is one kilogram of clean wool at the same micron worth

One micron finer What value is the next finest micron category in your style

One extra lamb How much income does a lamb contribute. Should include wool shorn and sale value of surplus animal.

4. The Target Levels of Production:

This information should indicate the levels of production the breeder wishes to be achieving in the future. For example, to increase clean fleece weight to 5 kgs and reduce micron to 21.5 in 10 years.

Class	Wool					Reproduction	
	Greasy Fleece Weight kg	Yield %	Micron	Style	Number of Joinings	% Lambs weaned per Ewe joined
Adult Ewes						
Wethers						

Environment:

Please describe the climatic and feed environment usually found.

Chapter 16

Breeding Objectives in Meat Sheep

Robert Banks

Introduction

Sheep meat industries across the world are facing severe problems arising mainly out of four factors:

- sheep carcases are small, resulting in lower processing efficiency compared to beef and pork,
- trading systems for all types of sheep are often not very objective, so that price signals do not flow readily from consumer to producer,
- lamb carcases particularly are perceived by potential consumers in higher-value markets as being over-fat. The small carcase size means that fat trimming is not an economically viable solution to this problem,
- sheep meat production is almost invariably a secondary enterprise to beef cattle, or wool production for example and far less effort is directed towards improving efficiency and profitability than in more economically important industries.

Better use of genetic resources offers considerable scope to tackle the biological components of these problems: doing so requires clear establishment of breeding objectives and their pursuit through efficient evaluation and improvement systems.

This Chapter outlines the breeding objectives in meat sheep production with a focus on lambs, and examines the ways in which industry structure can modify these objectives. The breeding objectives applied in the Australian lamb industry are outlined in some detail, followed by discussion of the situation in a number of countries that are significant producers of sheep meat.

Industry Structure in Australia

Australia produces some 17 million lamb carcases per year, worth about $Aus400 million. These carcases average 18 kg and 15 mm tissue depth at the GR site (110mm out from the backbone at the 12th rib). Of this production, some 17% is exported. In addition, some 16 million older lambs and mature sheep are slaughtered annually. Approximately 60% of these older sheep are exported, and the total value of mature sheep production is around $Aus250 million.

All the problems noted above affect the Australian sheep meats industry. Domestic consumption of lamb has declined steadily over the past 25 years, and consumer surveys highlight the problems of over-fatness and insufficient lean meat.

Australian sheep meat production is essentially a by-product of the wool industry, and produces animals for slaughter in the following classes:

- older or cast-for-age ewes and wethers (almost entirely Merinos),
- surplus young (around 12 months old) Merino ewes and wethers,
- first-cross lambs, produced in matings between long-wool breed (i.e. Border Leicester) rams and older Merino ewes,
- second-cross lambs, produced by mating first-cross ewes, mostly Border Leicester-Merino to terminal sire breed rams, mainly Poll Dorsets.

There is in addition some geographic structuring: Merinos tend to be run in areas of lower rainfall and hence shorter growing season for pastures; second-cross lambs are bred and grown in higher rainfall areas mainly in the south-eastern corner of Australia.

Finally, these production classes produce three types of carcases:

- lambs, sheep that have not yet developed their first adult teeth,
- hoggets, sheep with adult teeth and usually slaughtered at around 12 months of age,
- old sheep, which are anything from 2 to 7 (or more) years old.

Approximate prices (1992) for these animals are lambs $25-50, hoggets $10-20, and older sheep up to $15.

Hoggets and older sheep are almost entirely Merinos. The breed-type contributions to the lamb industry are:

Table 16.1 Breed type contribution to the lamb industry.

Breed Type	Proportion of genes in	
	Lambs	Dams of Lambs
Terminal sires	0.29	-
Crossing breeds	0.16	0.16
Self-replacing	0.21	0.27
Merino	0.31	0.58

NB: Crossing breeds are primarily Border Leicesters
The self-replacing breeds include Coopworths, Romneys, Corriedales, and the Carpet wool breeds.

Genetic improvement strategies for the sheep meat industry in Australia have so far concentrated on the terminal sire breeds and the breeds used in crossing with Merinos or in self-replacing flocks. Almost all selection effort in Merino flocks is directed towards improvement of wool production and quality, so that improvements in growth rate, carcase quality, and reproductive traits are very unlikely to derive from the Merino population.

Industry Use of Genetics

Two types of genetic variation are available for exploitation: between-breed variation, through breed substitution, cross-breeding, and new breed development, and within-breed variation, through selection programs.

The Australian prime lamb industry has traditionally been based on a biologically sensible cross-breeding structure: terminal sire breed rams are mated to either first-cross, usually Border Leicester-Merino or straight-bred such as Corriedale or Merino ewes.

This system exploits hybrid vigour for the maternal traits, reproduction and mothering ability where cross-bred ewes are used, and for growth traits because the prime lamb is a two- or three-way cross.

In theory, the system recognizes different breeding objectives for different component breeds: terminal sire breeds are selected for growth and muscular conformation, the dam breeds for reproductive ability and wool production, plus growth. How efficient these selections have been is open to some question.

Finally, Australia is currently importing Texels and American Suffolks, both of which are likely to be used as terminal sires, either as pure-breds or in synthetics. In addition Finnish Landrace are being imported, and may have a role in improving performance of maternal breeds, again through infusion. In general however, breed substitution through importation is not readily available to the Australian sheep meat industry because of very stringent quarantine regulations.

LAMBPLAN and Breeding Objectives in Meat Sheep

LAMBPLAN has been established to provide genetic information for the Australian sheep meat industry. The format of that information is designed with the production and breeding structure clearly in mind, and that format reflects breeding objectives for meat sheep breeders.

Firstly, in the terminal sire breeds, improvement of growth rate and leanness is desirable, and so for simplicity, the basic objective includes Weight and Subcutaneous Fat depth at 12 months. Because of the difficulty in pricing leanness, LAMBPLAN has taken an approach that offers different biological targets for breeders to aim at. These then are three standard objectives breeders may choose from:

HIGH GROWTH: places all emphasis on increasing growth rate

HIGH LEAN: places all emphasis on reducing subcutaneous fat depth

LEAN GROWTH: places equal emphasis on increasing growth rate or weight, and on reducing subcutaneous fat depth.

In addition, breeders may nominate their own combinations, and a small proportion of breeders are now using an option that places 75% emphasis on increasing weight and 25% on reducing subcutaneous fat depth.

The predicted genetic responses under these three objectives are:

Table 16.2 Predicted Genetic Parameters using LAMBPLAN

Selection Index	Emphasis Weight	Emphasis Fat	Predicted change over 10 years in (kgs)	(mm) @60kg
High Growth	+1	0	+10	0
High Lean	0	-1	+1	-4
Lean Growth	+1	-1	+7.5	-3
Breeders	+0.75	-0.25	+8.8	-1.5

These simple objectives have been readily adopted and understood by breeders. It is likely that if market signals improve and breeders obtain across-flock genetic evaluations, more price-based objectives will be used. So far, evidence available suggests that the LEAN GROWTH option provides a sensible approach to making satisfactory improvement in growth rate whilst simultaneously improving carcase composition.

Genetic evaluation of eye muscle area is being introduced to LAMBPLAN in 1992. Its value as a trait in the breeding objective is based on:

- direct improvement of muscle thickness is expected to improve visual appeal of many lamb cuts,
- research indicates that heavy lean lambs with large eye muscle area have carcases with a higher proportion if their lean tissue in the hind leg, chump and loin, which are the most valuable parts of the carcase.

As with fatness, imperfections in the market for lamb make objective pricing of eye muscle area and of carcase composition difficult at present in Australia. The approach being taken is an extension of that for weight and fat: a series on objective options which are best described as

172

biological targets and thus represent a desired-gains approach to formulating breeding objectives, incorporating weight, fat depth, and eye muscle area are provided to breeders.

The maternal sector of the lamb industry includes two broad groupings:

- crossing breeds, mainly Border Leicesters, mated with Merino ewes to produce F1 daughters,
- self-replacing breeds (Coopworths, Corriedales, carpet wool breeds etc), which as the name implies are to a large extent run in pure-bred flocks. However, these breeds are also used to a small extent in crossing with Merinos and with terminal sires.

The breeding objective for maternal breeds (Border Leicesters, Coopworths, Corriedales etc) is much more complicated for four reasons:

- more traits influence returns from the prime lamb dam - wool cut, fibre diameter in some breeds, reproductive rate, plus growth and leanness,
- there is considerable uncertainty about the overall value of increasing growth rate and hence increasing the mature size, and thus feed requirements, of ewes, and about the extent to which subcutaneous fat acts as an energy reservoir for ewes, and must be maintained to allow satisfactory reproductive performance.
- optimum lamb marking rates, and hence the desirability of selecting to increase reproductive rate, varies depending on geographical location, level of management expertise, and marketing options.
- the fibre diameter of wool strongly affects its value, and hence reduction of fibre diameter is apparently an important component of the objective for maternal breeds. In practice the most effective means of achieving this is to infuse Merino genetic material, so that breeders of maternal sheep usually either ignore fibre diameter, which has little effect in the 28-35 micron range, or change it rapidly by infusing Merinos.

In developing breeding objectives for the maternal sector, addressing these realities has lead to adoption of a simple desired-gains approach again. For each breed two options are provided:

- a formally derived objective using best available estimates of prices for lamb weight, fat depth, wool weight, fibre diameter, and value of extra lambs,
- a simple objective aiming at increasing weight, wool cut and reproductive rate, and placing zero economic weight on reduced fat depth and fibre diameter.

To date the second option has been more widely accepted by breeders.

Approaches to Breeding Objectives in other Countries

Breeding objectives applied in the Australian lamb industry reflect the biological and geographic structure of the industry, with clear separation of roles between terminal sire and maternal breeds. Where approaches differ in different countries they reflect different structures and different markets for different types of sheep meat.

This section briefly describes the approach to breeding objectives for sheep meat production in New Zealand, North America, and Western Europe. Clearly these areas are all western industrialised societies. Comment will also be made about objectives in sheep meat production in developing countries.

The New Zealand lamb industry differs from the Australian in two main ways: firstly it is highly export-oriented and this has encouraged development of more objective trading of lamb, and secondly it has been heavily based on use of self-replacing flocks, the main breeds being Romney and Coopworth. This second factor means that average carcase weights are lower than in Australia (around 14 kg), and that breeding objectives have aimed at improvement of both maternal phase (wool weight, reproduction) and growth phase (lamb weight, and more recently leanness) traits.

The predominance of self-replacing breeds continues, with only a small fraction (<10%) of lambs being sired by terminal sire breed rams. Recent importation of Texels from Scandinavia has not so far altered this situation, and it is quite likely that Texels will be infused into the self-replacing breeds to improve their carcase leanness and conformation.

The North American sheep industry makes use of grain-feeding to grow lambs and the lambs are grown out to heavier weights than in Australia or New Zealand. Carcase weights average 35 kg or greater, and the carcases are very fat, with fat trimming being widely used. The distinction between terminal sire and dual-purpose, self-replacing flocks is not as clear as in Australia, and the breeding objectives applied in all breeds are those relevant to dual-purpose animals: increased growth rate and improved maternal traits, particularly reproduction. Less attention is paid to leanness, wool weight or fibre diameter. One result of this approach is that North American strains of meat sheep breeds, for example Suffolks and Merinos are larger than those elsewhere.

In western Europe, most sheep meat production is based around dual-purpose breeds, with breeding objectives consequently aiming at improving both the growth and carcase traits and maternal performance. The degree of importance attached to carcase conformation depends heavily on the carcase grading system in use, which is evolving towards premiums for more muscular carcases. Most countries therefore use some moderate sized dual-purpose breed as the core of the industry, with some use of breeds with more muscular carcase shape, i.e. the Texel as a terminal sire.

In Great Britain, a stratified crossing system is used, with prime lambs predominantly sired by shortwool breeds (Suffolk, Dorset & Poll Dorset, Texel) rams over cross-bred ewes, often Leicester or Border Leicester crosses with Blackface and other hill breeds. The Australian system is very similar to this, except that Merinos fill the role of the hill breeds. Breeding objectives in flocks in Great Britain are therefore similar to those applied in Australia: growth and carcase merit in terminal sire breeds, maternal performance and wool cut in the dual-purpose breeds.

The countries and regions covered here all have to some degree organised breeding programs with some definition of breeding objectives. Sheep meat is clearly produced throughout large parts of the developing world. This production is largely from dual-purpose breeds, with much less application of objective breeding methods.

Conclusions

In the lamb industry of most developed countries sheep meats face problems, largely resulting from over-fatness and small carcase size. The Australian industry, whilst facing these problems, has the advantages of having a biologically sensible crossing system, and increasingly objective lamb trading. Use of the crossing system is reinforced by the market distinction between lamb and older sheep, and by the use of different climatic regions for producing F1 ewes and 3-way cross lambs.

The existence of this system reinforces clear distinction between terminal sire and dual-purpose crossing breeds, and provides scope for increasingly formal definition of breeding objectives. The brief survey of sheep meat production systems in other areas of the world highlights some important points about application of breeding objectives in meat sheep breeding:

- they are dependent on aspects of the marketing system: differentials between different ages, clarity of price signals for carcase attributes, and processing costs,
- they are dependent on the range of breeds available and the production system in which they are used,
- they are affected by environmental factors such as length of growing season, cost of feed, and scope for out-of-season lambing.

To sum up, the breeding objective for any objective meat sheep industry will include both growth phase and maternal phase traits. The extent to which these are combined in single objectives for a sheep population of a country or region depends mainly on the extent to which structured crossing is practised.

Chapter 17

Breeding Objectives for Dairy Cattle

Mike Goddard

Introduction

Milk production traits are clearly an important part of the breeding objective for dairy cattle but most countries publish sire evaluations for other traits and some cattle breeders regard these traits as very important. Emphasis placed on these other traits is often blamed for failure to achieve as much genetic progress in milk production as theoretically possible. However, the aim should be progress in overall profitability. To achieve the best balance between traits it is necessary to have a rational method of calculating the increase in profit from an improvement in each trait. This depends to some extent on the country and even the individual farm where these improvements are to be made. This chapter illustrates how economic weights can be derived for a typical Australian situation but the method should also be applicable elsewhere.

The selection index which is used to select bulls and cows is constrained by the information (e.g. EBVs) which is available. In the second part of this chapter, the traits for which EBVs are available in Australia are used to calculate an index. These results on selection objectives and indices for Australia are based on the work of Goddard (1985 and 1987), Beard (1988), Jones and Goddard (1991) and Malamaci (1991). Examples of selection objectives and indices for other countries can be found in Pearson (1982), and Niebel (1982).

Economic Weights

The first, and perhaps the most important, decision a dairy farmer must make concerning genetic improvement is the **direction** in which he wishes to change the herd. I will assume that this direction is that which maximises profit. In that case his aim when selecting a sire is to choose the sire with the most profitable daughters. To do this he needs to know how genetic change in traits such as milk yield affects profit. A profit function is a rule which relates genetic change to changes in profit. For instance, if a farmer is paid $4/kg for protein, $2/kg for fat and penalised 2.5 cents/kg for volume, a simple profit function might be

$$\text{Profit} = 4 \cdot \text{protein} + 2 \cdot \text{fat} - 0.025 \cdot \text{milk}$$

This simple rule illustrates several features of profit functions:

- The coefficient or economic weight for protein (i.e. 4) is the value of increasing protein when all other traits in the profit function are held constant. The negative economic weight for milk does not mean that low milk yield is desirable but that if fat

and protein are held constant a decrease in volume is beneficial. Similarly traits which only affect profit because they affect milk production need not be included. This is because when protein, fat and milk are held constant the extra trait has no effect on profit (i.e. its economic weight is zero). Thus only traits which directly affect profit need be included.
- The purpose of the profit function is not to predict farm profit but to show how a genetic **change** in each trait causes a **change** in profit. Thus traits in which there is no genetic variation can be ignored.

However, this simple profit function is inadequate because it ignores traits in which there is genetic variation such as milking speed, length of herd life and food intake. Food intake is particularly difficult to deal with. Our approach is to assume that

- the objective is to maximise profit from a farm of fixed size,
- the total amount of food consumed on the farm is not affected by the genetic merit of the cows (i.e. it is kept constant),
- the amount of food consumed per cow can by calculated from her bodyweight and milk production using standard feeding tables,
- because the total feed consumed is constant, changes in the requirements per cow are accommodated by changing the number of cows i.e. the stocking rate.

The sources of income included are sale of milk, cull cows and surplus calves. There are several costs which are unaffected by the genetic merit of the cows e.g. shire rates. These costs would simply subtract a constant from the profit function and so can be ignored. The main costs which need to be included are those which are proportional to the number of cows and calves i.e. herd costs, milking shed costs and calf rearing costs.

To calculate the effect of a genetic change on farm profit consider a standard herd of 150 milking cows, 32 yearlings and 35 calves. Now calculate the effect of a change in each trait on farm profit. For instance, a 1kg increase in milkfat yield per cow increases income by $2 per cow but the higher feed requirement decreases the number of cows that the farm can support. Consequently the net increase in annual farm profit is $130.

Similar calculations can be done for other traits and the results are given in Table 17.1.

Table 17.1 Economic weights for a breeding objective and selection index weights.

Trait	Economic Weights	Selection Index	Approximate Index
fat (kg)	130	65	1
protein (kg)	433	217	1
milk (l)	-6	-3	0
survival (%)	490	245	2
temperament (%)	161	161	2
milking speed (%)	134	134	1 or 2
overall type	240	120	1
bodyweight (kg)	-52		
size	0	-104	-1

The relative economic weights of protein, fat and milk are somewhat different to the 4:2:0.025 in the payment formula due to the feed required for each component. Fat contains more calories than protein and the food needed to synthesise lactose, which is in proportion to the volume, makes the weight for milk volume even more negative.

The length of herd life is measured as the proportion of cows which survive from one year to the next. In the standard herd 80% survive, so 20% are culled or die each year. Table 17.1 shows that a 1% increase in survival rate adds $490 to herd profit. This extra profit occurs because less calves must be raised as replacements and the herd contains less young cows which have a lower milk yield than mature cows.

Larger cows and their larger calves are worth more when sold but require more feed for maintenance. This necessitates a reduction in stocking rate. The net effect of higher meat value and reduced stocking rate is a reduction in farm profit of $52 for every 1kg increase in bodyweight (Table 17.1).

Temperament might not directly affect profit but is valued by dairy farmers for quality of life reasons. Wickham (1979) attempted to put an economic value on temperament by calculating the amount of milk yield that a farmer was willing to forgo in order to cull a cow for bad temperament. We used the same method.

Farmers in progeny test herds score heifers for temperament. Approximately 12% are scored as unsatisfactory. An increase in the proportion unsatisfactory to 13% was regarded by farmers as equivalent to a $161 loss of income from milk (Table 17.1).

Cows which are slow to milk increase total milking time. We valued milking speed by the cost of labour for this additional time. Farmers score heifers for milking speed in the same way as temperament. An increase in the proportion unsatisfactory for milking speed from 12% to 13% cost the farm $134 per year.

Type traits are considered valuable for 3 general reasons. Some like dairy character are correlated with milk yield. Some are correlated with length of herdlife or survival. Others may be of direct value in themselves. For instance, some farmers with right angle herring bone milking sheds find that it is difficult to put cups on cows with wide teat placement.

The first two classes of type traits need not be included in the breeding objective because when milk production and survival are held constant they have no additional value. The third category has a direct benefit to some farmers. This direct value varies among farmers and is difficult to assess. One way to set an economic weight for type traits is to use the method used for temperament i.e. farmers tendency to cull cows for type traits is compared with their tendency to cull for low production. The economic weight for overall type calculated by Malamaci (1991) using this method is given in Table 17.1.

Selection Criteria in Australia

Once the selection objective is decided the next step is to choose the selection criterion or index. The aim is to choose the index which is the best predictor of breeding value for the objective or profit. If all traits are included in a multi-trait analysis (e.g. in WOOLPLAN), the selection index is simply obtained by multiplying each EBV by its economic weight and adding them up. Dairy ABVs are calculated from single trait analyses but the same method of deriving an index is still a useful approximation. However allowance must be made for the fact that ABVs for production, type and survival are breeding values, while the ABVs for workability traits are transmitting abilities. This means, for instance, that one unit increase in ABV for milk yield produces a half unit increase in a bull's daughters, but a one unit increase in temperament ABV produces a one unit increase in his daughters. To account for this, the economic weights for production, type and survival have been halved to derive the selection index in Table 17.1.

Body weight was included in the selection objective but there is no ABV for bodyweight. Therefore the correlation between size and bodyweight has been used to derive an index weight for size ABVs. The weight is negative because, if other traits are held constant, big cows have higher maintenance requirements and so are less profitable.

The index weights in Table 17.1 should not be regarded as highly precise as a number of assumptions have been made in deriving them, and selection objectives will vary slightly from farm to farm. They should be regarded as an approximate guide.

Fortunately, when desirable traits are highly correlated, the index weights can be varied somewhat without a drastic decrease in the efficiency of the index. For instance, over a wide range of milk payment formulae, the selection index 'fat EBV + protein EBV' is 97% as efficient as the optimum combination of milk, fat and protein. Table 17.1 also contains an approximate, simplified index. This simplified index starts with 'fat EBV + protein EBV' (index weights of 1 and 1) and then expresses the other traits by the kg of 'fat + protein' to which they are economically equivalent. Thus a 1% increase in survival EBV is equivalent to 2 kg increase in fat + protein EBV.

Use of the simplified index to compare two bulls, Bozo and Glug, is illustrated in Table 17.2. Although Bozo is superior to Glug by only 5kg of fat + protein, when other traits are included Bozo is superior by the equivalent of 28kg. If you try this index on a number of bulls you will discover that milk production traits are by far the most important. A farmer who refuses to use a bull which is negative for type or below average for milking speed will reject many profitable bulls.

Table 17.2   Selection index calculations for two bulls.

Trait	Index Weight	Bozo EBV	Bozo Value	Glug EBV	Glug Value
fat + protein	1	60	60	55	55
survival	2	4	8	-2	-4
temperament	2	88	0	84	-8
milking speed	1	92	4	88	0
overall type	1	-2	-2	4	4
Size	-1	-2	2	3	-3
Total			72		44

Temperament and milking speed EBVs have been expressed as a deviation from average (88%) before calculating their contribution to the index

Value for Money in Semen

The selection indicies described above help dairy farmers to identify the best bulls, but in selecting which semen to purchase, they must also consider its price. The value per dose of semen of a 1kg increase in a bull's fat + protein EBV is approximately

$$\tfrac{1}{2} \cdot \tfrac{1}{5} \cdot 4 \cdot \$2.50 \cdot \tfrac{1}{2} = \$0.50$$

where

$\tfrac{1}{2}$ = proportion of a bull's superiority which is passed on to his daughters

$\tfrac{1}{5}$ = probability that an insemination produces a daughter that enters the milking herd

4 = number of lactations per cow (more strictly number of discounted expressions per cow)

$2.50 = value of 1kg of fat + protein made up of 0.57 kg fat and 0.43 kg protein

$\frac{1}{2}$ = the proportion of the extra income not absorbed by extra costs such as feed costs

That is, the dairy farmer can afford to pay **up to**, $0.50 extra per dose of semen for each additional kg of 'fat + protein' EBV.

This 50 cents/kg rule can be applied to the equivalent kg obtained by use of the simplified index. Since Bozo's index was 28kg better than Glug's, his semen is worth $14 per dose more. If Glug semen costs $20 per dose less than Bozo semen, it is the better buy.

References

Beard K (1988) Efficiency of index selection for dairy cattle using economic weights for major milk constituents. Aust J Agric Res 39:273-284

Goddard ME (1985) Breeding objectives for dairy cattle. Proc 5th Conf AAABG 248-253

Goddard ME (1987) Genetic Evaluation of dairy cattle for profitability. Proc 6th Conf AAABG 192-198

Jones LP and Goddard ME (1991) Value added bull selection. In 'Picking Winners' Goulburn Valley Dairy and Machinery Field Days, Stanhope

Malamaci K (1991) Fourth year BAgr Sci Thesis, LaTrobe University

Niebel E (1982) Economic evaluation of breeding objectives for intensive milk production in the temperate zone. Proc 3rd Wrld Congr Genet Appl Live Prod 9: 18-32

Pearson RE (1982) Economic evaluation of breeding objectives in dairy cattle. Proc 3rd Wrld Congr Genet Appl Live Prod 9: 11-17

Wickham BW (1979) Genetic parameters and economic values of traits other than production for dairy cattle. Proc NZ Soc Anim Prod 39:180-193

Chapter 18

Breeding Objectives in the Pig Industry

Tom Long

Direction and the Breeding Business

Many producers of breeding stock are: weighing pigs, probing pigs, counting numbers of piglets born per litter and making culling and selection decisions using this production information but have never taken a step back and critically evaluated what their breeding objective is. In designing effective breeding programs, determining the breeding objective is one of the most critical areas. Failure to critically address direction upfront and regularly thereafter does not auger well for business success. The breeding objective establishes the central direction for the breeding business.

There are several issues that should be considered when defining the breeding objective. The first issue to address is: where does the breeding business fit in with other enterprises the breeder might be involved in. Some breeders are also raising crops, cattle or sheep and the overall objective is to maximise profit from all these enterprises. Resources allocated to the pig breeding enterprise will depend on the other enterprises and this can affect the definition of the breeding objective in the pig enterprise. The next issue to consider is: what are the overall goals of the pig breeding enterprise?

Defining the Goal

When determining the overall goal of the pig breeding enterprise, several points need to be addressed.

- Does the breeder want to serve as a nucleus breeder or as a multiplier of the genetic improvement attained by other nucleus breeders?

  If the breeder wishes to be a multiplier then the breeding objective is much simpler than if the breeder desires to operate as a nucleus. The multiplier still has the goal of producing as many superior animals as possible for sale but the majority of the genetic improvement is coming from the nucleus where replacements are obtained from. Hence, a number of the decisions usually made in defining a breeding objective have been made in selecting a nucleus source of replacements.

- What market does the breeder want to supply with breeding stock?

  Commercial producers are utilising various crossing systems to produce their market pigs. Some are using a specific cross system which requires terminal sire boars and maternal line females. Other producers are using a 2 or 3 breed rotational system, where gilt replacements are produced on farm. This system requires 2 or 3 lines of pigs to be produced that are dual-purpose, i.e. need to have good performance for both production and reproductive traits. Other commercial producers are only using one line of pigs in their production units, a synthetic, and also require a dual-purpose line of pigs.

Breeders must determine which of the types of commercial producers they want to develop breeding stock for, as this will determine the resources needed for the pig breeding operation. If, for example, the breeder wanted to produce animals to be used in a specific cross, a terminal sire line could be developed to be used with another breeder's F1 females. Alternatively, the breeder could produce a complete package for the commercial producer, but this would require more commitment of resources as three lines of pigs would need to be maintained. Knowing what the markets are for breeding stock and being able to develop lines of pigs to meet those markets is key in developing a successful breeding operation. This is why defining breeding objectives is so important.

- What resources are available for the breeding program?

A critical constraint might be breeding herd size. A very small breeder should probably not try to breed a maternal line as inbreeding and genetic drift, coupled with low heritability for litter size, could frustrate endeavours to make genetic progress. However, this type of breeder could have some success in developing a terminal sire line. Inbreeding and drift would still be problems, but heritabilities would be more favourable to making genetic progress.

- What is the breeder's competitive position in the marketplace?

If, for example, a breeder had been selecting only for reduced backfat for a long period of time and had developed a very lean line of pigs (averaged 10mm of backfat on ad libitum feeding), he/she might want to change breeding objectives such that more selection pressure was applied to other traits of economic importance (e.g., growth). Their pigs might be known in the industry as being very lean and they would want to maintain that level of leanness in the line, but could redirect selection pressure to other traits of economic importance.

- Have present *versus* future gains been considered with regard to the profitability of the breeding program?

For example, rapid genetic progress could be made if only gilts were farrowed, thereby improving future gains, but present gains could be diminished due to the lower level of production of gilts relative to sows.

Traits Contributing to the Breeding Objective

In 1987 Barlow described defining the breeding objective as "Identifying the characteristics of the animals and herd or flock that contribute to change in profit, and their relative worth..."

To ascertain which characteristics of the animals contribute to a change in profit the breeder needs to determine important factors that contribute to costs and returns for the enterprise. The major costs to the production system are feed, housing and labour. These costs can be subdivided into costs in the growing-finishing sector of the system and costs in the gestation-farrowing sector of the system. Two breeding sub-objectives can be developed from this division, a growing-finishing breeding sub-objective and a sow breeding sub-objective. Admittedly these two sub-objectives are inter-related, but this partitioning makes definition of the breeding objective clearer. An example of traits which would impact on costs to the growing-finishing sub-objective would be growth rate and feed conversion efficiency. Genetic improvement in these traits would reduce costs to the growing-finishing sector. An example of a trait which would impact on costs to the sow sub-objective would be number of piglets born alive since genetic improvement in this trait would spread overhead costs per sow over a greater number of pigs. Returns must also be addressed as these contribute to profit. Price received for the market pig is the major source of return. Since most slaughter houses pay premiums for lean pigs and penalties for fat pigs, carcase yield and leanness would be important in the breeding objective.

Once traits in the breeding objective which contribute to profit have been identified, the economic worth of a unit change in each of these traits must be determined, i.e. what is the value in dollars of a change of 1 gram/day in Average Daily Gain (ADG). This can be a difficult task and can vary from one breeder to another. In deriving these economic values the breeder should consider an average value for his/her customers that is valid over the next 3-5 years as this is the target to breed for. Consideration of timing is very critical in establishing a successful breeding objective. The breeder also needs to choose selection criteria that are correlated with traits in the breeding objective, i.e. selecting for reduced P2 backfat measurements to improve carcase leanness at the slaughter house and improve returns. In addition to being correlated to traits in the breeding objective, these selection criteria should be relatively easy and inexpensive to measure. Finally, the breeder needs to combine selection criteria, economic values and traits in the breeding objective into a single value upon which to base selection, using genetic and phenotypic variances and covariances. This is what is done in developing a traditional selection index. An example of a procedure which can do this is the SELIND software package developed by Cunningham (1970). However, this procedure is not the optimal method for use in conjunction with modern genetic evaluation, see Chapter 11 on Genetic Evaluation in Pigs. A procedure which has been developed to accommodate breeding objectives within the BLUP methodology is the $INDEX module of the PIGBLUP system.

$INDEX

The $INDEX module, initially developed by Professor Terry Stewart (Stewart, *et al.* 1988), is a tool for pig breeders to use in helping them firmly establish their breeding objective. It uses a bio-economic profit function and budgeting approach. It operates by considering two sub-objectives of the pig's life cycle. The growing-finishing sub-objective uses economic and production inputs to define costs and returns in the growing-finishing department of the production unit so that the value of the trait EBVs for each pig can then be determined. Likewise, the sow sub-objective determines costs and returns for producing an additional pig so the EBV for number born alive (NBA) can be valued. These two sub-objectives are then combined into a single overall objective (the $EBV) upon which to base selection/culling decisions.

The $INDEX module uses economic and production inputs to determine the economic worth of each EBV for an animal, e.g. + 38 grams/day for ADG, by assessing costs and returns. For example, a boar with a high positive EBV for ADG will be expected to produce progeny which reach market weight sooner, thereby saving daily costs of maintaining a pig in the grower unit. Progeny from this boar would be more profitable as production costs would be reduced, assuming average EBV's for other traits. Likewise, a boar with a high positive EBV for backfat, a fat boar, will be expected to produce progeny that are fatter than average and, if there are penalties at the slaughter house for exceptionally fat pigs, these pigs would have carcases of reduced value. Therefore, progeny from this boar would be less profitable as returns would be reduced. For the sow sub-objective, a boar with a high EBV for NBA would be expected to have daughters with larger litters so would be more profitable than boars siring less productive sows. Using these costs and returns, the EBVs of each animal are combined into the respective sub-objectives, growing-finishing or sow. The economic and production inputs required by the system are presented in Tables 18.1 and 18.2.

Table 18.1 Economic Inputs to $INDEX

Item	Unit	Default	Breeder's Estimate
Average carcase market price	$/kg	2.15	-------
Premium for grid fat class cypher 0	$	0.00	-------
Premium for grid fat class cypher 1	$	0.00	-------
Premium for grid fat class cypher 2	$	0.00	-------
Premium for grid fat class cypher 3	$	0.00	-------
Premium for grid fat class cypher 4	$	0.00	-------
Premium for grid fat class cypher 5	$	0.00	-------
Cost of feed in the feeder	$/kg	0.24	-------
Non-feed costs per day	$/pig/day	0.15	-------

Once the EBVs of the animal have been combined into the growing-finishing and sow sub-objectives, these two sub-objectives are combined into a single total objective (the $EBV) for use in selection of replacements. The marketing inputs supplied by the breeder (Table 18.3) are used in combining these two sub-objectives. The marketing inputs should reflect percentages that are relevant to the breeder's herd. This enables the module to place the appropriate emphasis on production *versus* reproductive traits depending on the breeder's major market for breeding stock, i.e. terminal sires, breeding gilts or dual-purpose sires.

Table 18.2 Production Inputs to $INDEX

Item	Unit	Default	Breeder's Estimate
Number of pigs born alive	pigs	10.2	-------
Pre-weaning mortality	%	21	-------
Post-weaning mortality	%	1	-------
Average daily live weight gain	gm/day	517	-------
Average p2 fat depth	mm	13	-------
Live weight feed conversion	kg feed/kg gain	3.5	-------
Target carcase weight	kg	65	-------
Average dressing percentage	%	74	-------

Table 18.3 Marketing Inputs to $INDEX

Percentage of male pigs reared that are sold and/or used as:

Terminal sires (used to produce market pigs)	-------	%
Maternal sires (used to produce replacement gilts)	-------	%
Slaughter pigs	-------	%
Total	100	%

Percentage of all gilts reared that are sold and/or used as:

Replacement gilts	-------	%
Slaughter pigs	-------	%
Total	100	%

Example - Terminal Sire Line *vs* Maternal Line

Table 18.4 presents an example of economic and production inputs for a pig production unit. For this example, the assumption was made that average production levels would be the same for both a terminal sire and maternal line of pigs, which would probably not be true. Relative to backfat classes, penalties were given for animals in the fat class cyphers 3-5 (fatter pigs) with no premiums for pigs in cyphers 0-2 (leaner pigs). This example also used the same costs and prices for the two lines. Tables 18.5 presents the marketing inputs to $INDEX for the two lines. Note the difference in inputs for the two lines in gilts saved as replacement and the types of sires used to produce animals. Denoting these uses for animals allows the $INDEX module to weight the two breeding sub-objectives appropriately so that an overall breeding objective can be obtained.

Table 18.6 gives the Estimated Breeding Values (EBVs) produced by PIGBLUP and the $EBVs for 10 boars when the marketing inputs from Table 18.5 were used. The EBVs indicate the superiority, or inferiority, of each boar compared to the genetic base of the herd which is set at zero. These EBVs will be the same regardless of inputs to $INDEX. However, note the differences in rankings based on the $EBVs (boars are sorted on the terminal sire $EBV). Boar 786 is a superior boar with either specification. However, Boar 1196 is the 4th ranked boar on a terminal sire index but is ranked 8th on a maternal index due to the poor EBV for number born alive.

This example illustrates a tool for combining selection criteria, the EBVs, into a single value upon which to base selection. It combines the EBVs in a way customised for each breeder's situation, so that their individual breeding objectives can be addressed.

Regardless of the tool the breeder is using, traits that contribute to the breeding objective must be identified, economic emphasis or worth must be placed on those traits, and selection criteria found that are correlated to traits in the objective to maximise genetic gains in profit. Often defining breeding objectives is not done by the pig breeder or is done trivially, but this is a process which must be done in designing optimal breeding programs.

Table 18.4 Economic and Production Inputs to $INDEX

Average carcase market price	$/kg	2.15
Premium for grid fat class cypher 0	$	0.00
Premium for grid fat class cypher 1	$	0.00
Premium for grid fat class cypher 2	$	0.00
Premium for grid fat class cypher 3	$	-0.15
Premium for grid fat class cypher 4	$	-0.25
Premium for grid fat class cypher 5	$	-0.50
Cost of feed in the feeder	$/kg	0.24
Non-feed costs per day	$/pig/day	0.15
Number of pigs born alive	number	10.2
Pre-weaning mortality	%	21
Post-weaning mortality	%	1
Average daily live weight gain	gm/day	517
Average p2 fat depth	mm	13
Live weight feed conversion	kg/kg	3.5
Target carcase weight	kg	65
Average dressing percentage	%	74

Table 18.5 Marketing Inputs for Terminal Sire Line and Maternal Line

Marketing Inputs for Terminal Sire Line

Percent of boars sold (or used) as terminal sires	40%
Percent of boars sold (or used) as maternal sires	0%
Percent of boars sold (or used) as slaughter boars	60%
Total	100%
Percent of gilts sold (or used) as replacement gilts	5%
Percent of gilts sold (or used) as slaughter gilts	95%
Total	100%

Marketing Inputs for Maternal Line

Percent of boars sold (or used) as terminal sires	0%
Percent of boars sold (or used) as maternal sires	40%
Percent of boars sold (or used) as slaughter boars	60%
Total	100%
Percent of gilts sold (or used) as replacement gilts	50%
Percent of gilts sold (or used) as slaughter gilts	50%
Total	100%

Table 18.6 Estimated Breeding Values and $EBVs for Boars With Different Breeding Objectives

Boar	ADG	EBVs BF	NBA	Terminal Sire $EBV	Maternal $EBV
786	60	-0.51	0.33	103	87
272	45	-1.11	-0.09	78	61
782	51	0.87	0.16	69	57
1196	48	-0.50	-0.69	68	46
2991	34	-1.00	0.35	67	58
1297	41	-0.11	-0.10	64	50
2415	34	-0.86	0.09	63	51
470	34	-0.41	0.21	60	51
2764	35	-0.37	-0.23	56	42
580	47	0.95	-0.21	55	41

Guide for Consultants

Points to consider while consulting with a breeder regarding breeding objectives:

- Determine what part of the breeder's total operation is comprised by the pig breeding operation and resources available to it.
- Determine what sector, nucleus or multiplier the breeder wishes to operate in.
- Ascertain the type of market the breeder wants to produce breeding stock for, i.e. terminal sires, F1 females, synthetics.
- Establish the competitive position of the breeder in the marketplace, e.g., does he now have a lean line of pigs, female lines, etc..
- Determine the traits in the breeding objective.
- Compute the economic worth of traits in the breeding objective.
- Establish the selection criteria the breeder is prepared to record on each group of animals, which are correlated to traits in the breeding objective.
- Combine the above, using a software package such as SELIND.
- If using $INDEX, determine economic, production and marketing inputs relevant to breeder's situation and run $INDEX, preferably also examining the sensitivity of the results to a range of changes in input/output price relativities.
- Establish how the index should be implemented in the breeding program.

References

Barlow R (1987) An introduction to breeding objectives for livestock. Proc 6th Conf AAABG, Perth, 9-11 Feb: pp 162-169

Brascamp, EW, deVries AG (1992) Defining the breeding goals for pig improvement. Pig News and Information, 13:21N-26N

Cunningham EP (1970) SELIND User's Guide

deVries AG (1989) A method to incorporate competitive position in the breeding goal. Anim Prod 48:221-227

Nicholas F (1988) In "Veterinary Genetics" ed. pp.479-482

Stewart TS, Harris DL, Boche DH (1988) A profit function approach to developing selection criteria in swine. Proc 7th Conf AAABG, Armidale, 26-29 Sept: pp 126-135

Stewart TS, Boche DH, Harris DL, Einstein ME, Lofgren DL, Schinckel AP (1990) A bioeconomic profit function for swine production: application to developing optimal multitrait selection indexes. J Anim Breed Genet 107:340-350

PART IV: Design of Breeding Programs

Chapter 19

Principles of Genetic Progress.

Brian Kinghorn

Manipulating Genetic Differences

Genetic resources constitute the engines of all food production systems:

- Plants and animals convert inputs to outputs.
- Better plants and animals do the job more efficiently.
- We can improve plants and animals by changing them **genetically**.

The animal breeder has two main tools for changing animals genetically - selection and crossbreeding. These are both passively implemented in that the breeder invokes change by choice of parents and allocation of mates. The transmission of genetic material takes place naturally at the genetic level, despite any use of artificial insemination or embryo transfer at the reproductive level, and the breeder plays only a guiding role.

On the other hand, active implementation of genetic change can involve, for example, laboratory manufacture of recombinant DNA constructs with microinjection into early embryos. This approach has yet to realise commercial viability, and it is not covered in this book.

This chapter concerns the principles of genetic progress as achieved by within-breed selection policies. Selection programs aim to change animals genetically in order to do a better job. We are aiming to create new strains genetically superior to what we currently have - not simply equal to the best we currently have. In chapters 4 and 5 the basis was given for predicting the merit of progeny out of a selected mating. In this chapter we show how to predict the rate of genetic change from a policy of ongoing selection.

Simple Selection Theory

First consider the relationship between offspring merit (O) and the mean merit of unselected parental pairs ($\frac{1}{2}P_m + \frac{1}{2}P_f = \bar{P}$, m for male and f for female):

One selection policy is represented by only using parents with $\bar{P}$ above a truncation point T, as shown in Figure 19.1. The average $\bar{P}$ of these selected animals is S units above the overall mean. S is called the Selection Differential, and it describes the mean observed superiority of parents. It should be noted that S is increased by allocating mates after selecting parents, as is normal practice.

The average value of progeny out of these selected parents is R units above the average of progeny out of unselected parents. R is the response to this single round of selection (i.e. response per generation), and as the regression slope of O on $\bar{P}$ is h^2, we have:

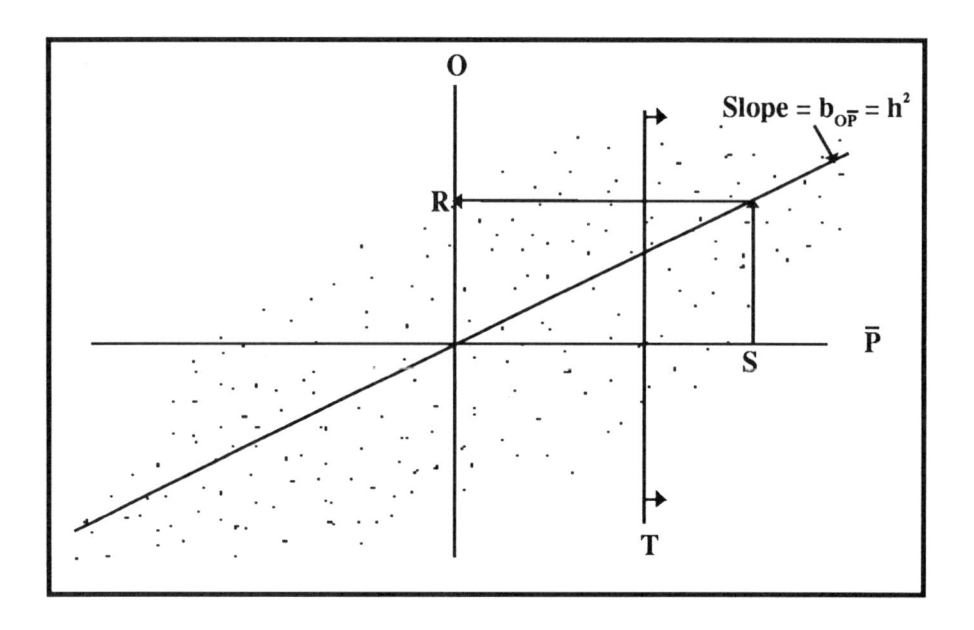

Fig. 19.1 The relationship between offspring merit (O) and mean merit of the two parents ($\bar{P}$). T is the truncation point above which pairs might have been chosen in a selection program, and S, the selection differential, is the average superiority of these selected parents. R is the response to selection. This is the average merit of offspring out of selected parents, expressed as a difference from what the merit on offspring out of unselected or randomly selected parents would have been. The slope of the regression line $b_{O\bar{P}}$ equals heritability, h^2.

> The regression slope of O on $\bar{P}$ turns out to equal heritability. Skip the contents of this box if you are prepared to believe that!
>
> $$\hat{O} = b_{O\bar{P}} \cdot \bar{P} \quad \ldots 1.$$
>
> and $\quad \hat{O} = \frac{1}{2}(\hat{A}_m + \hat{A}_f) \quad$ ....from chapter 4
>
> so $\quad \frac{1}{2}(\hat{A}_m + \hat{A}_f) = b_{O\bar{P}} \frac{1}{2}(P_m + P_f) \quad$ ....from 1. above
>
> $$\hat{A}_m + \hat{A}_f = b_{O\bar{P}} P_m + b_{O\bar{P}} P_f$$
>
> and as $\quad \hat{A} = h^2 P \quad$ we get $b_{O\bar{P}} = h^2$

A Definition of Heritability:

Heritability is the regression of offspring on parental mean.

Or in more digestable terms,

Heritability is the efficiency of transmission of parental phenotypic superiority to the next generation.

Each sex is equally important in contributing genes to the next generation, so calculate the selection differential separately for each sex, then average them:

$$S = \frac{S_m + S_f}{2}$$

This can also be seen graphically in Figure 19.2. The scale of these distributions is observed phenotype, not EBV as chapter 4.

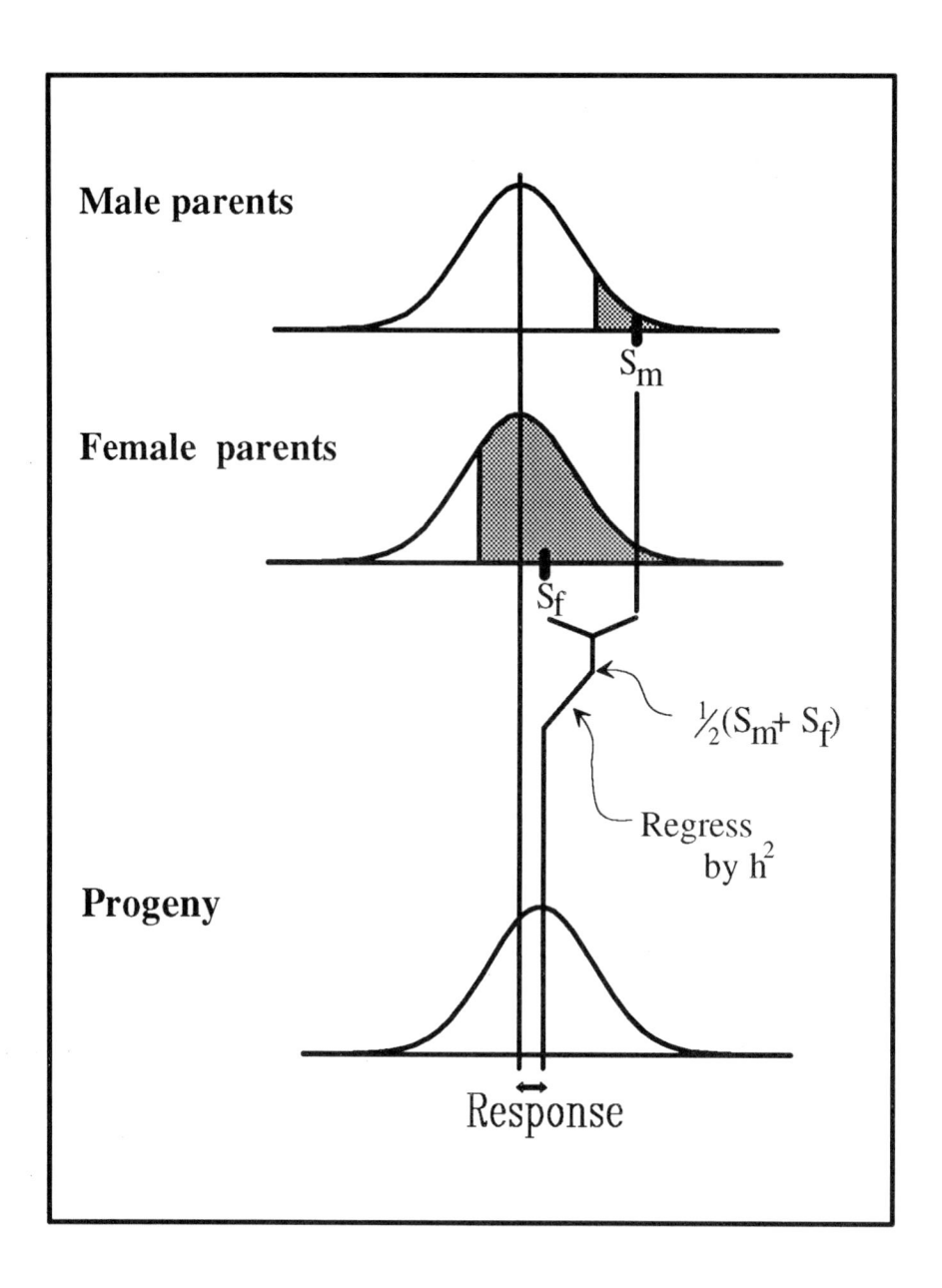

Fig. 19.2 An illustration of selection differentials for each sex and the resulting response in the next generation. Notice higher fecundity in males means that fewer need to be selected, giving a higher selection differential for males. The overall selection differential is regressed by heritability to predict response. In chapter 4, EBVs were not regressed - they are pre-regressed and therefore fully heritable.

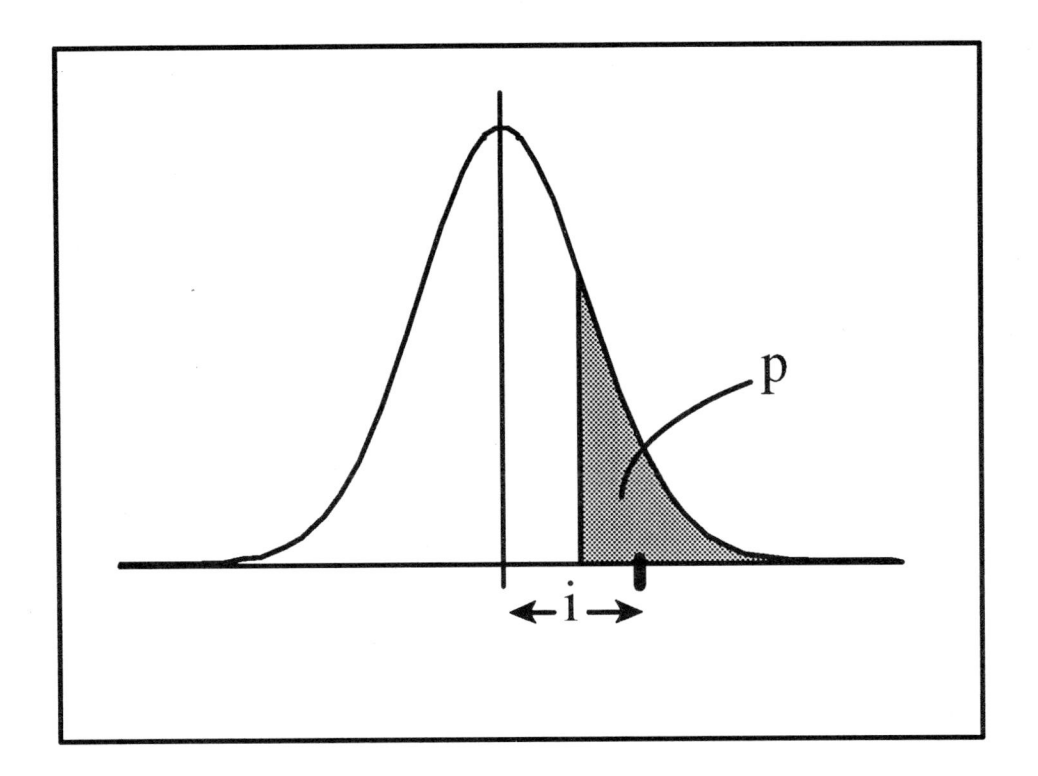

Fig. 19.3 Selection intensity (i) is a description of the expected average superiority of selected animals, as a function of the proportion selected (p). Selection intensity has units of phenotypic standard deviations, and tables or computer algorithms are use to find them, given proportions selected.

Selection Intensity

We want to be able to predict response without having to actually measure selection differential, S. We tackle this by predicting S itself from a knowledge of the proportion of animals we intend to retain for breeding (p) and assuming normality, as illustrated in Figure 19.3. Therefore selection intensity i is the number of standard deviation units (σ) that selected parents are superior to the mean.

Proportion Selected	Selection Intensity
p	i
0.01	2.7
0.02	2.4
0.05	2.1
0.10	1.8
0.25	1.3
0.50	0.8
0.75	0.4
0.95	0.1

Selecting the cream / Selecting the dregs

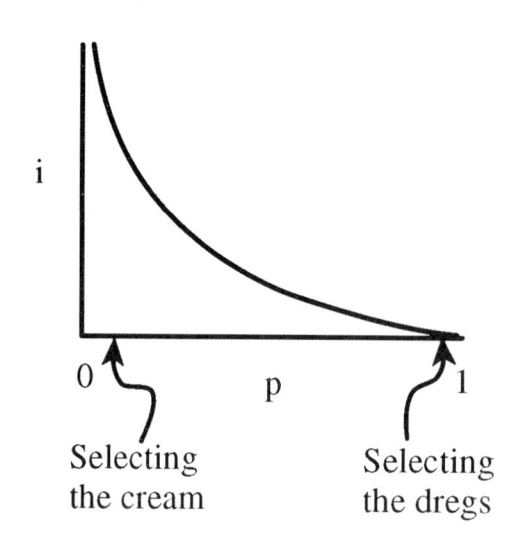

Fig. 19.4 Selection intensity as a function of proportion selected. Note the trend that as fewer animals are selected a greater intensity of selection is achieved.

Just as we average selection differentials to get an overall selection differential, we can calculate an overall selection intensity in the same manner:

$$S = \tfrac{1}{2}(S_m + S_f) \quad \text{and} \quad i = \tfrac{1}{2}(i_m + i_f)$$

Now we can predict the selection differential without the effort of making actual measurements on candidates for selection:

$$\text{Predicted Selection Differential} = \hat{S} = i\sigma$$

And as $R = h^2 S$, we can also predict response per generation to selection:

$$\text{Predicted Response per generation} = \hat{R} = h^2 i \sigma$$

In these equations, the ^ (indicating 'estimate') is conventionally dropped from S and R. Figure 19.5 gives an example of calculating selection insensity, selection differential and response to selection.

> **EXAMPLE** - yearling weight in beef cattle
>
> Average yearling weight of 100 bulls: $\bar{X}_m$ = 300kg, $\bar{X}_f$ = 275kg, h^2 = 0.25, σ = 30kg
>
> **What is the expected average weight of the top 10 bulls?**
>
> $S_m = i\sigma p$ $p_m = 10/100$ giving $i_m = 1.755$, σ = 30kg
>
> $S_m = 1.755 \times 30 = +52.65$kg superior, over 300.   Answer = 352.65kg
>
> **What is the response to selecting these bulls over random cows ?**
>
> $R = \frac{1}{2}(i_m + i_f) h^2 \sigma$ - where i_f is expected to be zero
>
> $R = \frac{1}{2}(1.755 + 0) \cdot 0.25 \cdot 30 = 6.58$kg response
>
> Note: This gives an expectation of 300 + 6.58kg for male progeny and 275 + 6.58kg for female progeny.
>
> **What is the response to selecting these bulls over the best half of the heifers ?**
>
> $p_f = \frac{1}{2}$ giving $i_f = 0.798$
>
> $R = i h^2 \sigma = \frac{1}{2}(1.755 + 0.798) \cdot 0.25 \cdot 30 = 9.57$kg response
>
> Note: This gives an expectation of 300 + 9.57kg for male progeny and 275 + 9.57kg for female progeny.

Fig. 19.5   Example calculations of selection insensity, selection differential and response to selection.

In chapter 4, the effect on offspring performance of using an individual animal was looked at, whereas in this chapter increased merit through a selection policy has been discussed. These two sources of genetic improvement can be mixed and matched as shown in Figure 19.6.

Consideration	Response	=	Male contribution	+	Female contribution
Individuals	$\hat{G}_O$	=	$\frac{1}{2} A_m$	+	$\frac{1}{2} A_f$
	$\hat{G}_O$	=	$\frac{1}{2} h^2 P_m$	+	$\frac{1}{2} h^2 P_f$
Groups	R	=	$\frac{1}{2} h^2 S_m$	+	$\frac{1}{2} h^2 S_f$
	R	=	$\frac{1}{2} i_m h^2 \sigma$	+	$\frac{1}{2} i_f h^2 \sigma$

EXAMPLE: As for Figure 19.5 but use one single 352.65kg bull over the best half (group) of heifers: Response = R = $\frac{1}{2} h^2 P_m + \frac{1}{2} i_f h^2 \sigma$ = 6.58 + 2.99 + 9.57kg

Fig. 19.6 Response to selection can be calculated for choice of individual animals and group selection policy, separately for sex

Selection Based on other Sources of Information.

In the last section response to selection on animals' phenotypes was predicted per generation as $R = i h^2 \sigma$. This equation can be re-arranged:

$$R = i h^2 \sigma = i \frac{V_A}{V_P} \sigma = i \frac{\sigma_A}{\sigma} \sigma_A = i h \sigma_A$$

- where σ_A is the standard deviation of (true) breeding values, **h** is the accuracy of selection, which is in fact the correlation between true and estimated breeding value, $r_{A\hat{A}}$ as introduced in the last talk. We now have an equation to predict selection response using EBVs calculated from any source of information:

$$R \quad = \quad i \quad\quad r_{A\hat{A}} \quad\quad \sigma_A$$

Response = intensity · accuracy · spread

For example, if selection of animals is based on the mean of n of their progeny, then selction accuracy is

$$r_{A\hat{A}} = \sqrt{\frac{n}{n + \frac{4-h^2}{h^2}}}$$

This would only be valid where both male and female candidates have n progeny measured. A more flexible approach is to treat each sex separately, accounting for the different selection intensities and selection accuracies in each sex:

$$R = \frac{i_m\, r_{A\hat{A}_m} + i_f\, r_{A\hat{A}_f}}{2}\, \sigma_A$$

This is a prediction of response per generation. The generation interval is the average age of parents when their progeny are born, with equal weighting to each sex. So the generation interval is $L = \frac{L_m + L_f}{2}$, where L_m is the average age of male parents when their progeny are born. As an example, the generation interval for sheep is usually about 3.25 years. This consideration leads to a prediction of response per year:

$$R = \frac{i_m\, r_{A\hat{A}_m} + i_f\, r_{A\hat{A}_f}}{L_m + L_f}\, \sigma_A$$

Lowly fecund species need to be bred for many years in order to provide sufficient candidates for replacement as breeeders. Thus generation intervals are typically about 4 years for beef cattle, 3.25 years for sheep, 2 years for pigs and possibly less than a year for poultry.

Genetic Progress in Open Nucleus schemes.

Breeding units which supply all their own replacement breeders can be referred to as closed nuclei. They export rams and ewes to other flocks, referred to here as base flocks, in order to disseminate their genetic gains. Figure 19.7 shows the selection and migration of ewes in such a nucleus scheme, using distributions of dollar EBV in the nucleus and base flocks or units. The dark shaded ewes are selected for use in the nucleus, and the light shaded ewes, some born in the nucleus, are selected for use in the base. It can be seen that the base contains some ewes which are of higher EBV than some of the less meritous ewes from the nucleus.

Open nucleus schemes exploit these competitive animals (mostly ewes, but sometimes rams) which arise in the base units or flocks, as seen in Figure 19.8. The average EBV of the ewes selected for use in the nucleus is now considerably more. This makes for faster genetic progress in the nucleus, with flow-on effects to the whole population.

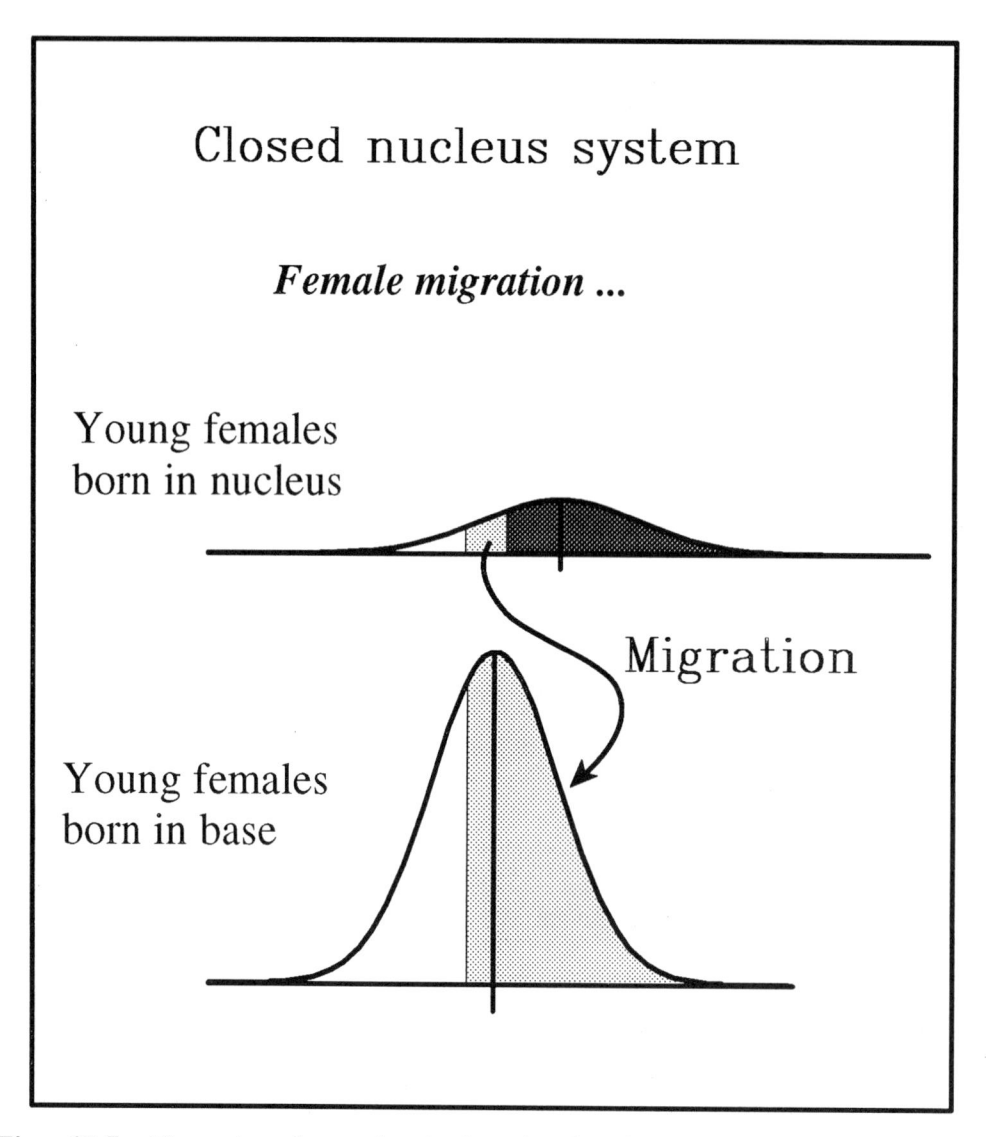

Fig. 19.7 The pattern of ewe migration in a closed nucleus sheep breeding scheme. The scale is in units of Estimated Breeding Value (EBV) such that animals in the base can be compared for breeding merit on the same scale as those born in the nucleus. The dark shaded females are selected for use in the nucleus, and the light shaded females, some born in the nucleus, are slected for use in the base. Notice that some base born females are better than some of those selected in the nucleus.

John James has shown that these benefits are about 10%-15% extra rate of gain. For three-tier schemes, the value of an open policy is worth about 22%. These figures assume, however, that animals in lower tiers are evaluated as accurately as in upper tiers. Of course it is wise to place greater evaluation effort in the higher tiers, and as such the percentage benefit from opening the system is somewhat lower.

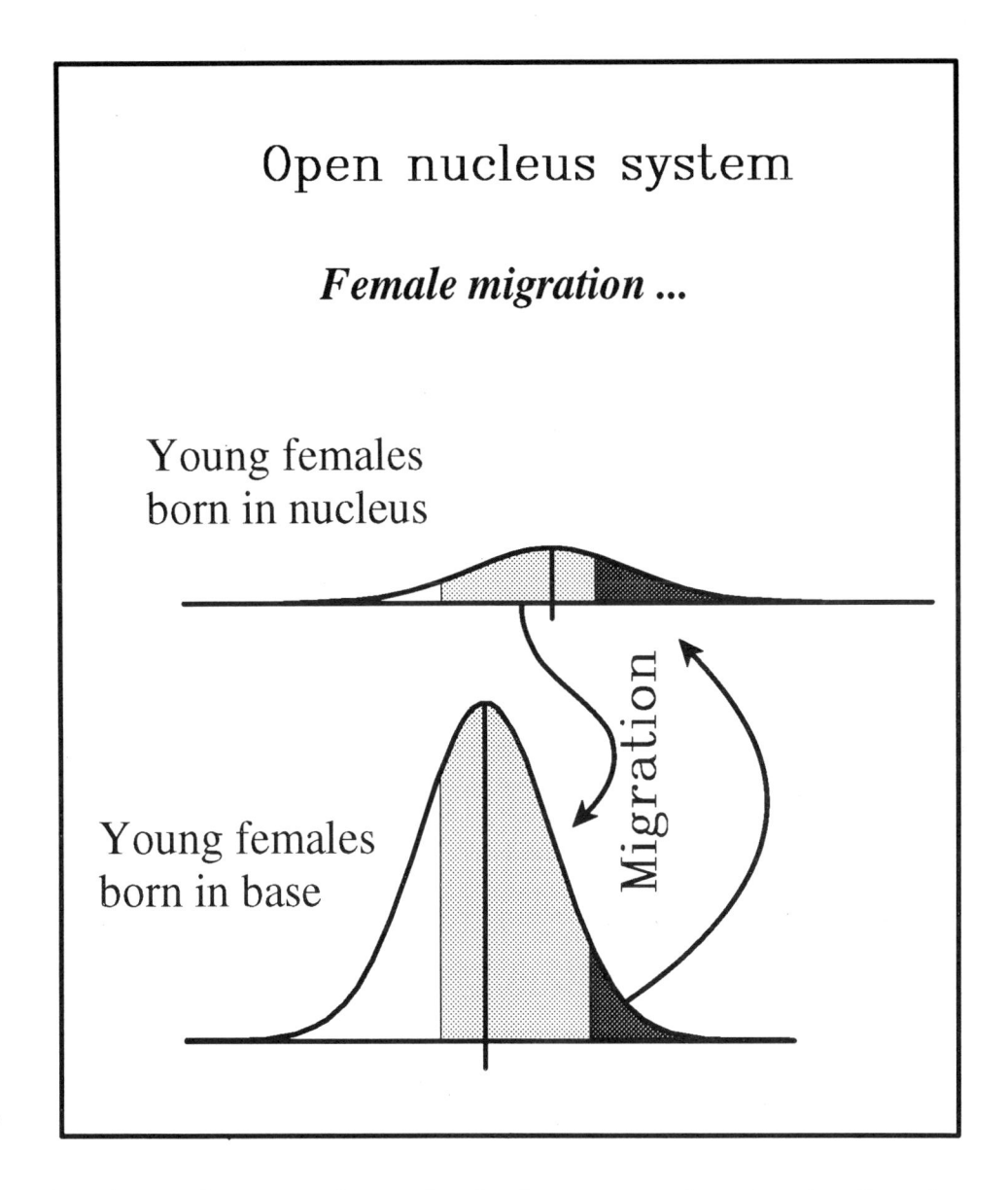

Fig. 19.8 The pattern of ewe migration in an open nucleus sheep breeding scheme. Compared to Figure 19.7, competitive females from the base are migrated to the nucleus. This gives faster progress in the nucleus, and ultimately in the whole scheme.

References

James JW (1977) Open nucleus breeding systems. Anim Prod 24: 287-305

Chapter 20

Design of Straight-Breeding Programs - Common Problems

Keith Hammond

Ideally, we may wish to **simultaneously** manipulate via selection, culling and mating **all** genetic differences of interest, i.e. differences at the single gene level, additive and non-additive genetic differences between breeds and crosses and within breeds, for both direct and indirect, e.g. maternal traits and, for repeated traits and permanent environmental differences.

The ideal world is not achievable, at least in the '90s Currently, we more or less sequentially operate on the single gene *vs* the within-breed quantitative variation *vs* the between-breed and between-cross genetic differences.

Sequential manipulation or not, it is still all about definition ($ direction), selection, genetic evaluation, mating structure, design, **and** about arranging the breeding operations to maintain profitability.

The formal aspects of design are overlaid by the operational or practical aspects at the management level of the herd or flock.

Operationally, at every point it is critical to ask: **What is the impact of this management decision or action on the effectiveness of the breeding program?** There are many, many links in the breeding chain. Secondly: **How can this decision or action be modified, within the context of the total management of the herd or flock, to achieve greater genetic gains and/or reduce the costs of the breeding operation and otherwise increase profitability?**

What are the Available Options for a Selection Program?
And what are the pros and cons of each?

The options are:

- Close the herd.
- Purchase all replacement breeding stock.
- Breed some and purchase some.

Close the herd and breed all your own replacement and sale stock. One of the greatest assets of the Australian livestock industries, in terms of achieving continual improvement, is its large herds and flocks. With a well-planned and efficiently managed program of genetic improvement beef herds of 150 cows or more can remain closed for 50 years or more and show continual improvement. These breeders have the advantage of being able to control their direction of improvement and of having access to the best rather than the second best replacements.

However, breeders rarely use this option, for two reasons. Firstly, the grass often looks greener on the other side of the fence. They think the stock of one or more other breeders are better than their own. Unfortunately, it is often only the grass that is better - the stock look better because they have grown under more favourable conditions, not because they have better genes. Often these stock will be purchased on no more than a pedigree. Remember two things: the eventual consumers of the industrys' products do not eat pedigrees; and you cannot look directly up an animal's genes for the four most important groups of characters to the industry - reproduction, mothering ability, growth, fibre and carcase and meat quality attributes. Measurements must be included in your decision making. Secondly, the shrewd breeder recognises the power of public relations - to buy male breeding stock, often for exorbitant prices, with stud prefixes which are currently popular and to pepper a sale with their offspring may mean additional returns. For some time this procedure became an end in itself rather than being combined with performance recording and carefully planned programs of genetic improvement; however it is now commencing to disappear.

Purchase all the replacement breeding stock. Usually only males are involved but some breeders also purchase the majority of their female replacements. In using this option, the breeder may purchase males or semen. Artificial Insemination programs require careful management but a number of situations exist where there are real advantages in using semen. For example, in small herds and flocks AI can help maintain a good mix of genes and hybrid vigour or, when setting up a herd or flock, it can be used to rapidly incorporate a wide sample of genetic material so the subsequent genetic improvement program will be more successful.

Of course the breeder who purchases his replacements or the majority of them, is not really a breeder but a **multiplier** of genetic material which has been bred by others. In this respect the number of genuine breeders in Australia is likely to be quite small.

Breed some and purchase some of your replacement stock. The majority of stud breeders use this option. The extent of purchase and use of outside males is generally high enough to mean that the most genetic progress or regress occurring in the herd is coming from outside. If the herds or flocks from which you are buying males are genuinely improving genetically for the characters of prime importance to you then your herd will improve, but it will always lag behind your seedstock suppliers by about 5 to 10 years. If the males you purchase are from herds or flocks which are marking time genetically then yours will also be static.

The purchaser of breeding stock must be fully aware of a basic principle on which our economy revolves: "Let the buyer beware"! When buying stock this principle must be applied particularly to:

- The seller's breeding objective and selection criteria?
- The nature and efficiency of the seller's breeding program - are the stock equal or superior to stock from other sources, and is the genetic potential of your herd likely to continue to increase over time if you continue to use this source?

- How the information provided is being obtained and presented for the stock available for sale?
- How you obtain valid comparisons between herds?

How Far Ahead to Plan?

At least 2 to 3 generations, i.e. 8 - 12 years for cattle. Difficult? Yes. Impossible? No. The true breeder has no option; even to stand still requires definition and action!

Key Points in the Design of Selection Programs

The following applies to all seed-stock producers - the registered and the non-registered or commercial breeders. It applies whether you are commencing a selection program or having a second look.

Definition

- Consider your breeding objectives and selection index carefully (explicit is preferable to implicit definition), remembering the environment you are producing in and the slaughter and seed-stock markets for which you are producing.
- Include all economic traits in the breeding objective, but keep the selection index simple.
- Resist changes in fashion - there is room for diverse types but breed for your environment.

Herd size

- Start with a larger gene pool.
- Close the herd if it is large enough - why not breed your own replacement males?
- Maximise the number of sires - foundation and annually.
- Make further introductions only after convincing yourself of their superiority in relation to **your** objectives.
- Small herds or flocks can make continual use of external artificial breeding, **if** the semen is available - **remember** from where the progress is coming - but use a good number of males.
- Take special care when contemplating the use of new techniques in reproduction - inbreeding depression and chance sampling (genetic drift).

Selection:

Accuracy.
- Keep good records, on the whole drop where possible.
- Select within **commonly** managed **large** groups.
- Maximise the number of sires represented in each management group.
- Minimise spread of calving and adjust records for biases.

Amount
- Concentrate on your selection index and be ruthless.
- Select replacement sires and dams from the whole drop if possible.
- Reduce joining percentage.
- Selection is increased with more joinings per year, e.g. 2 in cattle.
- Maximise branding percentage.
- Increase herd size.

Generation turnover
- Use males as young as possible - their genes do not improve with age!
- In using BREEDPLAN type genetic evaluation procedures for all traits of interest replace sires only with superior males unless inbreeding depression is becoming a concern.
- Otherwise, turn males over rapidly - regardless of a male's superiority, some of his sons will be even better, so continue to use the 'new models'.

Inbreeding depression
- Use more than 5 sires per year *to reduce impact of genetic drift more than inbreeding*.
- Turn sires over rapidly.
- Small herds or flocks (6-10 sires used per year) - select no more than 2 replacement males per sire progeny group.
- Avoid more than 1 common grandparent in a mating in small herds or flocks.
- When multiple-sire joining in cattle, use the first half of each multiple-sire bull group for 3 weeks, and the second half for the remainder of the joining.
- Increase the joining percentage.
- Consider closing herds or flocks which use more than 10 new sires per year - inbreeding is of no importance, unless you are involved in a sound across-herd/flock genetic evaluation scheme.

Mating
- Consider single-sire mating, but effective programs are certainly possible with multiple-sire mating (Kerr *et al.* 1992).
- Of course mating best to best increases your chance of producing an outstanding offspring.

Records
- All genetic improvement relies on comparing performance of different animals or groups.
- Choose your system(s) of animal identification carefully.
- Consider carefully your method of collecting and keeping records.

Integrating the selection program and management system

- Aim for **simplicity** - your program will be more effective.
- **Do not fiddle** - continuing progress requires steady pressure in the chosen direction on your whole gene pool.
- Design your program into your management system, **not** the reverse.
- Use pencil and paper to draw up **every** step in one complete cycle of your program then proceed to make refinements and talk the **plan** over with appropriate advisors.

Common Problems In Selection Programs

The breeding program which cannot be made more efficient and effective does not exist!

Appreciable improvements to the design and/or operation of all existing programs could be achieved by altering one or more of the following for the herd or flock. These are not in order of importance:

- Careful **definition** of breeding objectives for the herd and of the selection criteria on which replacement decisions are to be made. This certainly means writing these down completely - to be referred to, and maybe updated, once every two to five years and so prevent the mind from wandering (we are all fallible!), or even creating a tread-mill situation and achieving no change at all.
- Maximise the **size of**:
  - The **breeding herd or flock,** and
  - The **groups** from which replacements are selected, being careful to maximise the number of sires represented in each management group.

    This will help you achieve as much selection pressure as possible and hedge against the influence of chance (luck) going against your attempts to improve. In addition, the larger the breeding herd the less the potential problems of inbreeding. **Do not castrate large numbers** of male offspring at branding/marking.
- **Guard against over-emphasising accuracy** of selection at the expense of amount of selection and generation interval.
- With the current genetic evaluation procedures not including all traits likely to be of importance, the **length of time which sires are kept is frequently too long.** Turn them over - their genes do not improve with age and, if you are improving your herd, the best of their sons should be superior to them. Use them for one or two years only. Of course, by keeping sires longer the amount of selection pressure can be increased i.e. less replacement sires are required per year, but the resulting increase in the generation interval generally more than soaks up any additional improvement from the extra selection pressure.

- The **average age of the female herd or flock** can be decreased by increasing the annual replacement rate, reducing the generation interval on the female side. However, this is done at the expense of slightly reducing the amount of selection pressure, because of the higher number of heifer, hogget or gilt replacements and increasing the running cost of females. This is generally not a major area for action, except in very old herds or flocks.
- Of greater importance is **the age groups of female breeders from which replacements are sourced.** The failure to consider offspring from the young age groups of breeders, including heifers, hoggets or gilts!, when selecting replacements increases the generation interval and reduces the amount of selection which can be applied because there will be less offspring from which to select. If the herd or flock is improving, the younger age groups should be superior!
- The **ratio of males to females** varies between herds or flocks from less than 1:30 to more than 1:60. The less males required the greater the potential for selection but in small herds or flocks (150 to 300) the number of males used per year will also have to take into account inbreeding and chance effects. It is better to use a couple more males per year in a small herd or flock to hedge against these two potential problems.
- The **level of knowledge** on all aspects of breeding in the industry must be improved. A concerted effort is required by all sectors associated with the industry and by the industry itself.
- Associated with the previous point, many breeders lack **confidence** in their ability to further concentrate the superior genes in their gene pool.

Labour Requirements And Design of Selection Programs

The total labour requirement of the herd or flock is **a major variable cost** and a breeding program places additional demands on the nature or type and quality of this labour, its management and the amount required.

There are a number of **operations associated with the breeding program** which can be varied to change the type, amount and quality of labour required and the effectiveness of the program. These are summarised in Figure 20.1. Examine each question in the Figure in fine detail for each breeding program.

Given a set of staff and its standard of management, there are a number of aspects of the breeding program which can be varied to change the total amount of labour required, the major aspects being: identification, recording, and managing stock.

- **Identification of stock**
  The **system of identification used should vary** with:
    - The intensity of identification, i.e. whether groups of animals (herds, grades, age or sire groups) or individual animals are being identified.

- Accuracy and permanency required.

All existing methods and systems (combinations of methods) are deficient in at least two of the important properties - legibility, speed and ease of application, permanency, safety, price, capability of use in yards and paddocks.

Consider carefully what animals you need to identify, at what age and the coding system you will use.

- **Collecting, maintaining and using records**

The total recording process is labour intensive and very careful thought and **planning** should be given to establishing a recording system for your herd or flock which has the essential properties of being:

- Simple
- Sufficiently accurate.
- Up to date at all times.
- Easy to retrieve information from.

A herd's recording system will comprise two basic parts:

- Paddock and yard procedures.
- Office procedures, which may involve:
  - A permanent, and
  - A temporary system of recording.

The final recording system used must be tailored to the individual property and herd or flock management procedure. The use of a day-book by all staff should be part of all on-farm herd recording systems.

Be realistic when evaluating the costs and benefits of changes to your recording system. Some small investments have the potential to achieve appreciable savings or returns from the breeding herd or flock.

It is **vital that time be allocated** in the daily and weekly work schedule for recording purposes, particularly for the office procedure as this is often the first to suffer. Recording is not a spare time job!

- **Management of stock**

Many aspects of herd or flock management require additional demands on labour when continual genetic improvement in a herd or flock is being sought. For example:

- The number of potential breeders for replacement and sale will be greater and they will need to be kept for a longer period.
- The additional movement of stock for purposes of recording, culling and selection, and joining, and the greater number of groups into which the herd is divided may very substantially increase the annual labour requirement.

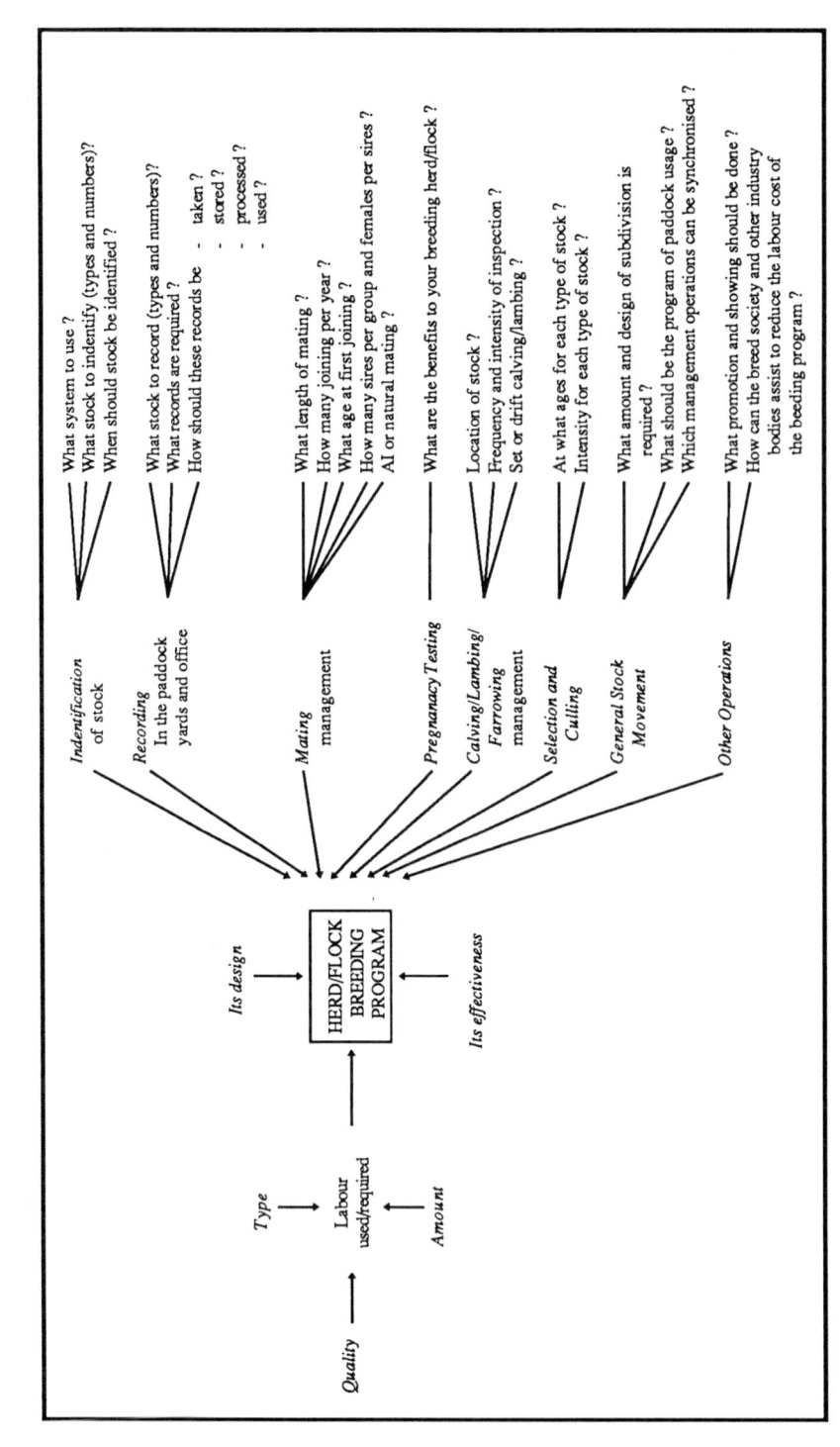

Figure 20.1 Operations contributing to the differences between herds or flock in the type, amount and labour used in breeding programs and influencing the effectiveness of the breeding program for your herd(s)

- **Minimise these additional requirements for labour** by
  - Careful design of the genetic improvement program, particularly in terms of:
    - The characters recorded and used to make selection and culling decisions,
    - The number of mating/calving or lambing groups - costly subdivision for single-sire joining may not be warranted,
    - The age at which male and female stock are first mated/sold, and
    - The length of joining and the number of joinings per year.
  - Developing the 'mini-care' concept in your herd or flock.
  - Co-ordinating stock movement and other stock management requirements.
  - Utilising techniques which facilitate and minimise stock movement - for example: lanes; electric fencing, particularly for bull management; decentralised yards; good roads.

Where Are The Payoffs?

- Cheaper replacements.
- More product of higher quality:
  - from genetic progress - the 'hinge'.
  - management windfalls - the 'gate'.
- Satisfaction from your own program.
- Creation of a capital asset.

Single Genes In The Breeding Program

Of the 100,000 or so genes contained in the 30 pairs of chromosomes in cattle only a very small number, currently about 50, have effects on the animal that are individually identifiable and are of economic importance. These are the ones which code for either:

- An obviously abnormal condition or defect; or
- A condition considered to produce a more saleable animal, for example polledness (several genes involved!) or various coat colours.

Where the economic incentive is large enough we can conduct a breeding program to reduce the occurrence or completely remove the unwanted form of a particular gene from the herd and increase or fix the favoured form, so that it alone remains.

The speed at which we change the relative occurrence of each form of one of these genes in the herd or flock and the design of the breeding program required will depend on the way the alternative forms of the gene behave with respect to each other; that is, on the type of inheritance involved. The common types of inheritance for single genes are complete dominance, partial dominance and sex-linked inheritance.

The breeding program for single gene traits

Certainly for the **simply inherited** single gene traits, breeding programs aimed at concentrating the desired form of the gene in the herd can be simply designed. For single gene traits that are not so simply inherited a little more trouble, and often expense, is involved in the design of the breeding program.

We briefly consider alternatives when the type of inheritance is complete dominance - the most common.

We may wish either to concentrate a form of a gene (red coat colour), or eliminate the deleterious form of a gene (dwarfism).

These objectives are really the same. In the first case we wish to identify and breed from those animals which possess two doses of the favoured form, be it the dominant or the recessive. In the second case, we wish to eliminate stock which show the defect and the carrier animals. If the defect is determined by the dominant form of the gene, then life is made easy - eliminate everything showing the condition. However, the majority of simply inherited defects in cattle are produced by the recessive form of the gene - the carrier now looks normal.

Consequently the majority of situations call for Test Mating i.e. the need to produce a number of progeny to detect the carriers. In years to come this will be overcome for some conditions by the development of biochemical tests to detect the carriers, tests which can be used early in the life of the animal, saving much time and money. Current examples are Mannosidosis in Angus, Galloways and Murray Greys, Pompes Disease (Glycogenesis) in Brahams and Beef Shorthorns, Citrullanaemia in Holsteins and the Halothane reaction in pigs.

Various forms of test mating can be used, each of which has advantages and disadvantages. For most forms of test, the number of fertile matings, assuming all progeny are observed, will determine the chance that a correct decision is made. Consequently, all we need to do is decide on the chance we are prepared to take, that is a correct decision 75% or 90% or 95% or even 99% of the time, and we can calculate the number of matings required to test a sire or dam.

For this type of inheritance the first calf born exhibiting the unwanted condition proves that the sire is a carrier **but** a sire can never be completely proved to be a non-carrier. Obviously the more normal progeny he has the closer to zero is the chance that he is a carrier.

The disadvantages of making special test matings are appreciable. For example, they take time to produce results, particularly when using a sire's own daughters. In addition, they tie up appreciable numbers of animals. Sometimes these are difficult to obtain but at all times special test matings require the maintenance of a special group of breeding females (the affected or carrier or test daughters).

To overcome the problems with special test matings which we noted above, we could simply join a male to a random group of females in the herd where the allele is present. The frequency of occurrence of the defect in the herd or flock now influences the number of matings required to detect a carrier.

In special cases, i.e. where the economic incentive is large enough, the numbers of females needed for special test matings can be reduced by the use of superovulation and embryo transfer. Where the defect being tested for can be identified in foetuses e.g. mule foot in 60-day foetuses, the testing program can be speeded up by removing and examining the foetuses.

Introducing an allele: When wishing to introduce a form of a gene that you currently do not have in the herd there are two basic sources:

- The result of a mutation.
- From another breed or strain.

Individual genes mutate once in every 100,000 to 100,000,000 million animals depending on the particular gene, that is, extremely rarely. Consequently, the only reasonable source of the desired form of the gene is another strain or breed.

The actual design of the breeding program will depend on the type of inheritance of the form of the gene required (whether recessive: red coat colour; or dominant: polledness) and the existence of other complexities, e.g. scurs create some problems in breeding for polledness. However, the breeding program will involve the use of back-crossing (grading-up) to the original strain while retaining the new form of the gene, test mating to detect the carriers, and selection to use the best animals for all other traits of interest as replacement breeders.

The great problem of trying to introduce a new form of a gene is that much selection pressure on the other traits of economic importance is foregone while fixing the new form of the gene in your herd, which will take at least 4-5 generations to accomplish. Consequently it should only be tried in large herds or flocks.

Record keeping

It is absolutely essential to obtain the best diagnosis possible of every defective offspring dropped and to keep a good record of it with its pedigree. Why? Because the proportion of carriers in your herd or flock will generally be so much greater than the frequency of affected offspring. The breeder who hides defectives 'under the carpet' and keeps no records of them is a fool. The chances are the condition will increase insidiously through carriers until it becomes a major problem. This also applies to the breed as a whole.

References

Kerr RJ, Hammond K, Kinghorn BP (1992) Efficiency of multiple-sire mating. Proc 10th Conf, AAABG, Rockhapmton, [in press]

Chapter 21

Maximising Improvement with AI, MOET and Cloning.

Brian Kinghorn

Introduction

Recall from chapter 4 that EBVs calculated properly from a selection index calculation or a BLUP (Best Linear Unbiased Prediction) analysis have one very useful property: The predicted merit of progeny is simply the average of the EBVs of the two parents used, as shown in Figure 21.1. A smaller proportion of males than females can be selected for breeding, contributing to their high mean EBV. The other factor, in this case, is the greater amount of information used to calculate male EBVs, reflected by a wider distribution of EBVs. Notice that the predicted merit of progeny is simply the average, or half-way-point between the selected males' and females' average EBVs. The width of the EBV distribution of the progeny depends on how intensively they are measured.

Improvements in Genetic Gain with A.I. and M.O.E.T.

Artificial insemination (AI) and multiple ovulation and embryo transfer (MOET, or just ET) serve to increase the prolificacy of chosen males and females respectively. This increase in prolificacy, if properly exploited, can bring two types of benefit:

- increased selection intensity, and
- more information for estimating breeding value.

Increased selection intensity

With use of AI, the few best males can be selected for breeding. This means that the average EBV of males used is higher. This can be seen in Figure 21.2. Only the very best males contribute to the average EBV of males. This increase is diluted 50 percent by the female contribution, but the net increase in predicted merit of progeny is quite visible compared to Figure 21.1 with lower selection intensity of males due to use of natural mating.

M.O.E.T.: Increased selection intensity and more information for estimating breeding value

With MOET, each selected female can contribute not just one or two offspring, but up to about 5 per donor female, a figure which continues to improve. As with AI, this brings about the ability to select fewer females *as donors of genetic material*.

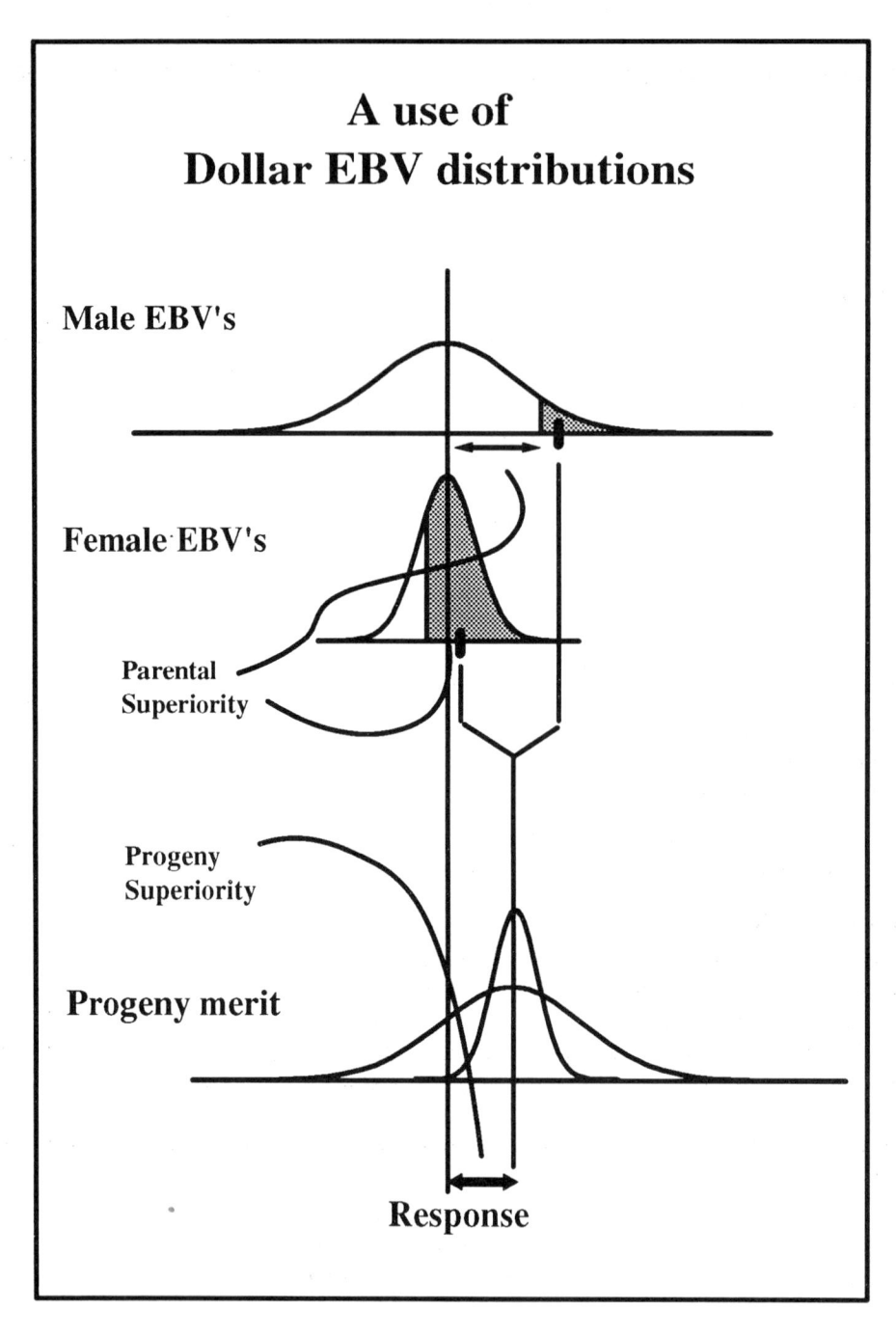

Fig. 21.1 The predicted response in progeny is simply the average of the parental EBVs

Many recipient females are still required to carry the offspring. Moreover, in an ongoing breeding program using MOET, candidates for selection will usually have a number of full-brothers and full-sisters available with records. This information helps to improve the accuracy of EBVs, and the distribution of EBVs increases accordingly.

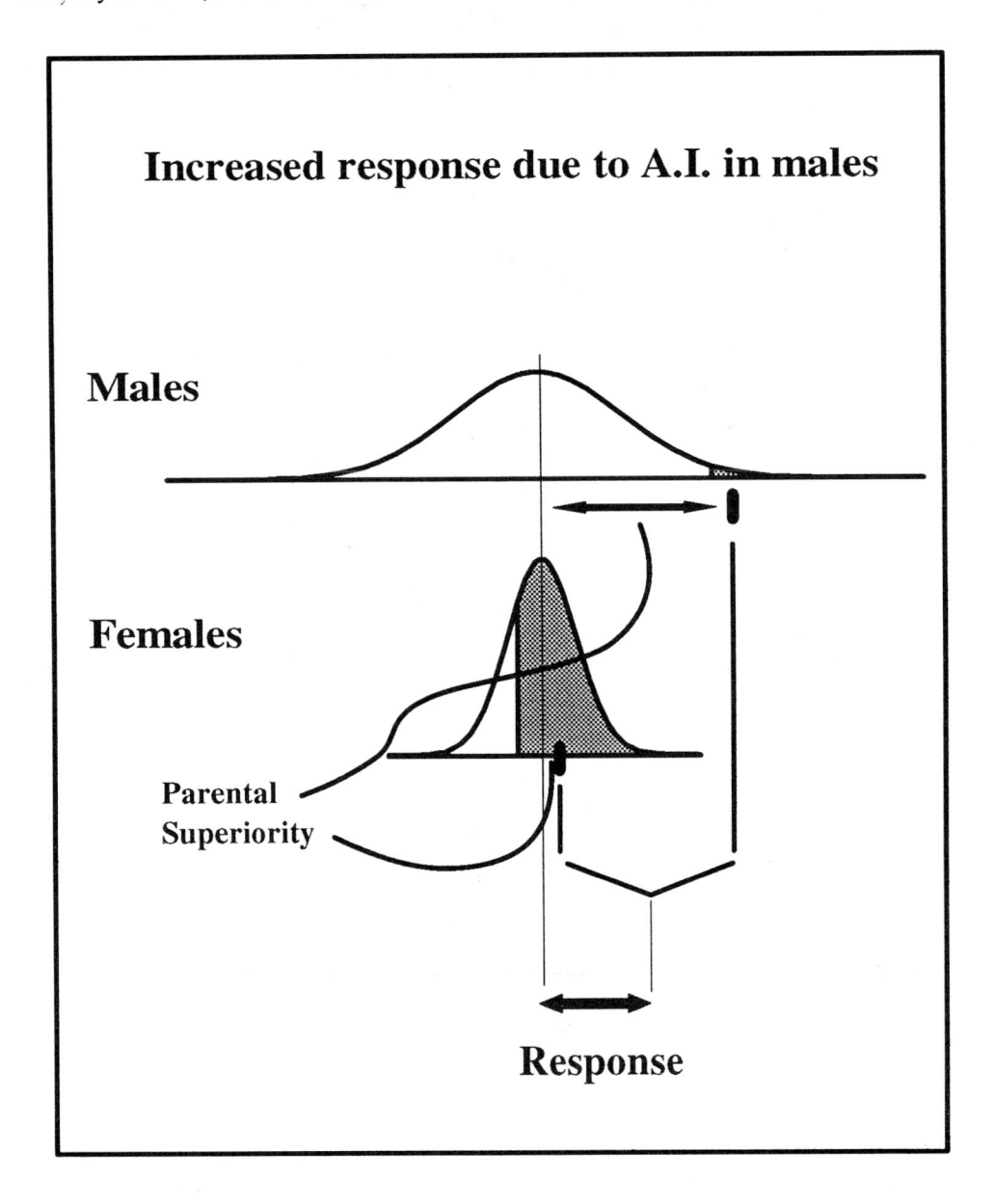

Fig. 21.2 The ability to select fewer males when using artificial insemination leads to higher selection intensity and more response. The scale is in units of EBV.

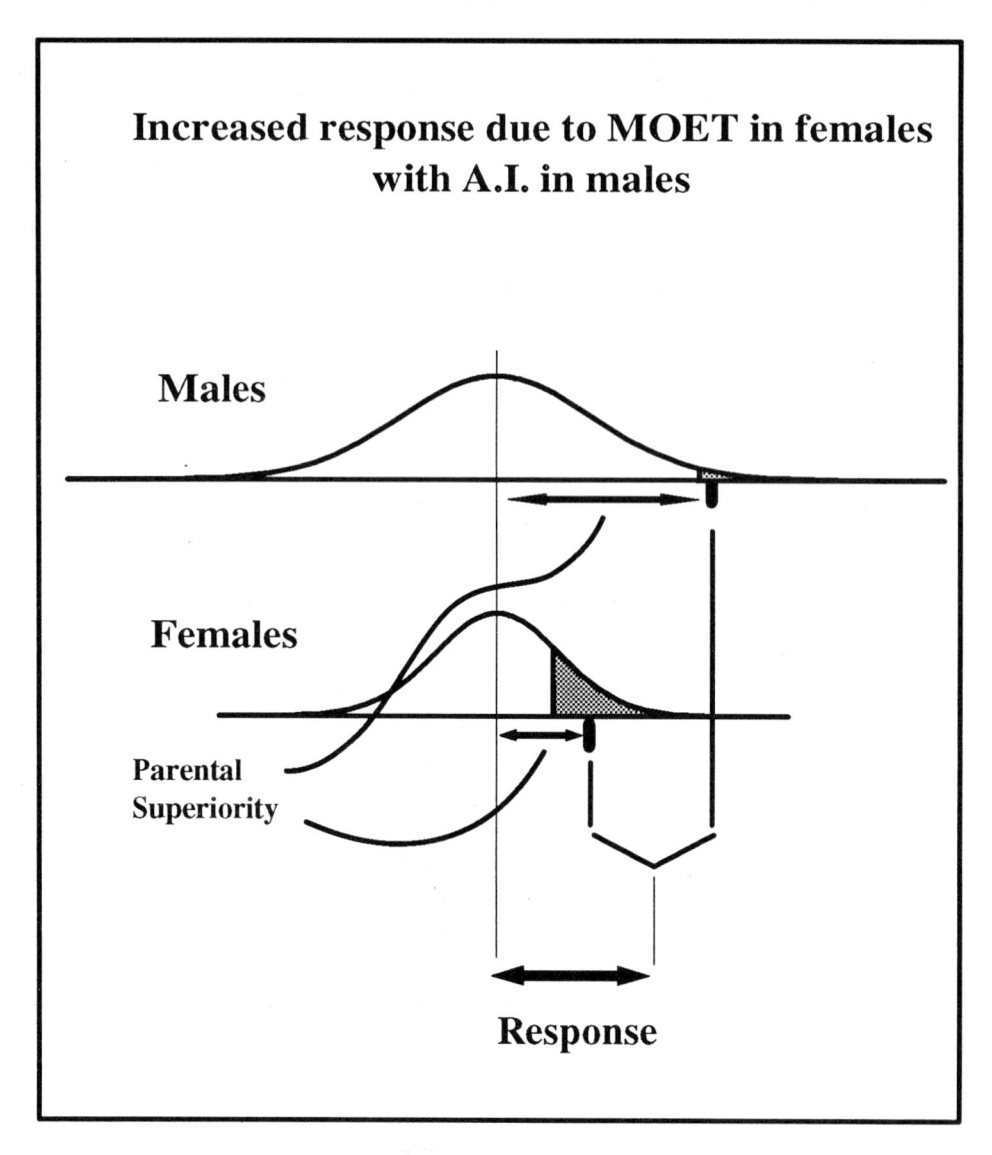

Increased response due to MOET in females with A.I. in males

Males

Females

Parental Superiority

Response

Fig. 21.3 MOET of breeding females also leads to higher selection intensity, but the increased numbers of full sibs per family gives more accurate genetic evaluation and a wider spread in EBVs. These effects both lead to more response.

Both these favourable effects are seen in Figure 21.3. Comparing this figure to the original natural mating figure (21.1), it can be seen that AI and MOET, if properly used, can bring about a notable increase in the response to selection. Simple theoretical predictions suggest that a MOET program will give about 50% extra gain over a normal breeding program.

This figure reduces to about 20-30% when account is taken of finite population effects and variance loss due to selection. In particular, care is needed to design the mating structure to manage the potentially rapid build up of inbreeding. Keller *et al.* (1990) and Kinghorn *et al.* (1991) discuss the issues of variance loss and inbreeding.

Improvements due to cloning.

We can now generate small numbers of genetically identical individuals from embryonic material. We cannot know that the resulting animals will be genetically elite. We can only aim at this by using elite parents, and bad luck can still play a role in the choice of individuals and individual embryos to clone.

It has been suggested that groups of such clones could be used to give accurate evaluation of themselves. Ten identical clones give ten estimates of the genetic merit of the one genotype involved. However, this tests the value of an animal's genes to itself, and this can be quite different from the value of an animal's genes to its progeny. The latter appropriately measured in a progeny test or BLUP analysis. Both this and the reduced number of different genotypes which can be tested, place some doubt on the value of using clones to help in genetic evaluation.

Of course this does not preclude wide testing of clones to find elite genotypes to use directly for commercial production, rather than breeding. This might be of most interest for commercially important traits, such as meat quality, which are difficult to measure as they require high labour costs or the need to slaughter animals.

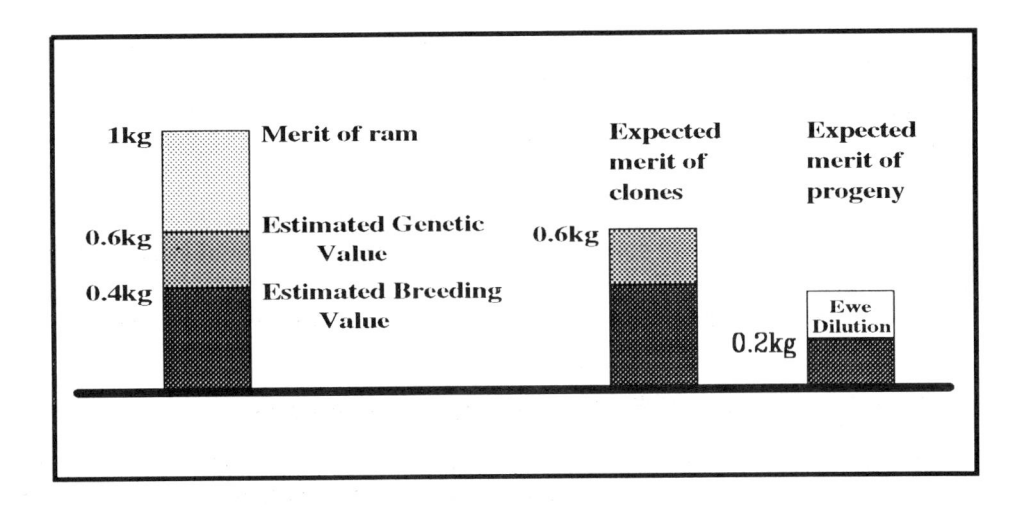

Fig. 21.4 Merit of clones and merit of progeny from a ram with a 1 kg superiority in fleece weight.

We cannot yet make identical clones of mature animals - animals which we know to be of exceptional observed merit for commercially important traits. This has been done in frogs, but not in mammals. How useful would this process be? Figure 21.4 makes a comparison of exploiting a superior ram either through cloning or generating progeny, either naturally or by AI. First of all, some of the ram's observed superiority of 1kg fleece weight is expected to be due to luck in the environmental factors which have affected his performance. This accounts for about 40% of his performance at the top of the bar in the diagram.

About 40% (the heritability of fleece weight) of his performance is expected to be reflected by the value of his genes to his progeny, giving an EBV based on his own performance of 0.4kg. However, favourable interactions among his own genes are expected to result in about 60% of his superiority being due to the value of his own genes to himself. This is also the expected value of his genes to his identical clones, such that the expected superiority of his clones is not 1kg, but 0.6kg.

If the ram is mated to average ewes, the performance of his progeny is expected to be the average of his EBV (0.4kg) and that of average ewes (0kg superiority), or 0.2kg. Thus cloning would be expected to yield about 3 times as much extra merit (0.6kg versus 0.2kg). Whether we could spread an individuals genes as widely through cloning as we can through AI remains to be seen.

IF we were able to make clones from mature individuals, and if we could disperse them widely, what would be the effect on average genetic merit in our livestock populations? Figure 21.5 shows progress of normal animals in the breeding program, and elite clones performing considerably better in any one year. After how many years will normal animals be performing as well as today's elite clones? This can be calculated making some simple assumptions: For example, for sheep:

	Normal Breeding Program	Clone Selection
Selection intensity	1.4	3.0
Heritability	0.4	0.6
Generation Intensity	3.25	3.25

The calculated answer is that today's elite clones are expected to be as good as normal sheep born in just over 10 years time. One estimate puts the likely cost of clone transfer at between $30 and $40 per lamb born. This would have to decrease substantially to make the 10 year jump worthwhile, especially as ongoing investment is required to keep ahead.

Use of A.I. and M.O.E.T. in open nucleus schemes.

Why is it that we do not migrate males as well as females from lower tiers to upper tiers in open nucleus breeding schemes? The reason is that the selection intensity in males is so high that base males can scarcely compete. This is especially the case with AI, as such a small proportion of available males needs to be selected, virtually all of these will consistently lie in the right-hand tail of the nucleus EBV distribution. Moreover, the expense of recording all males in the base just to send a few to the nucleus may be very hard to justify.

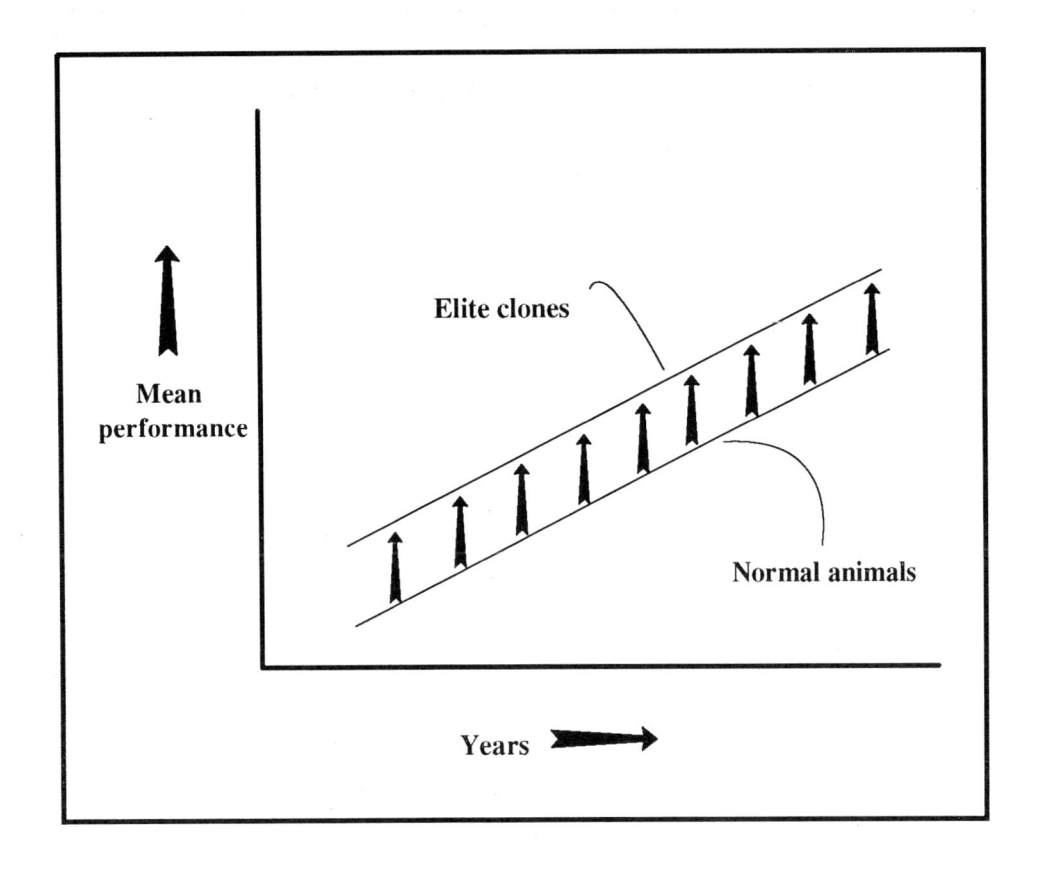

Fig. 21.5 Genetic progress in a normal breeding program, and in elite clones derived from it.

There is a similar story when the nucleus herd or flock consists of MOET donor females. In order to provide the same numbers of nucleus progeny, a MOET scheme requires perhaps only a half to a third of the number of females, depending on the effectiveness of the MOET program. This makes it more difficult for base-born females to compete - such that a lower proportion should be migrated upwards. This effect may be aided by increased accuracy of EBVs for nucleus-born females, due to the extra information available from their full-sisters and full-brothers. Both these effects contribute to the low level of upward migration shown for an open MOET-nucleus scheme in Figure 21.6. Comparing Figure 21.6 with Figure 19.8, which has no MOET of nucleus ewes, shows the large effect of MOET on optimal migration rates.

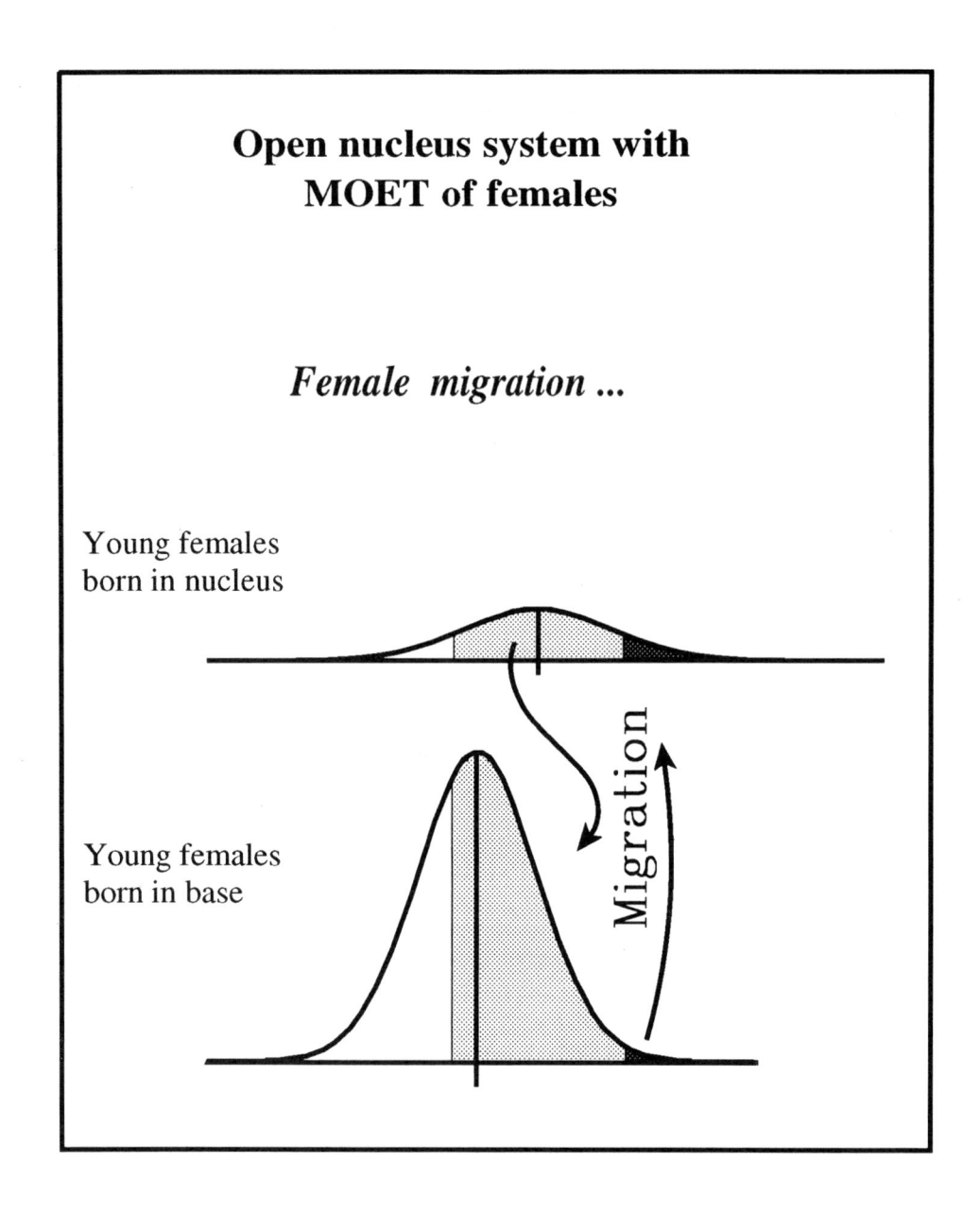

Fig. 21.6 Migration of females in an open nucleus system with MOET of nucleus breeding females.

Dairy cattle progeny testing programs can be considered as open nucleus systems, with elite females identified in a number of herds for contract matings with top sires in a geographically dispersed nucleus. MOET of these elite females is now proving popular and valuable. However, with the change of fecundity imposed by MOET, radically different structures, which can exploit the lower generation intervals made possible, show good potential (Nicholas and Smith, 1983). With the advent of female gamete harvesting and in vitro fertilisation, even more different structures are indicated, and predicted rates of gain become very high (Kinghorn *et al.* 1991).

References

Keller DS, Gearheart WW and Smith C (1990) A comparison of factors reducing selection response in closed nucleus breeding schemes. J Anim Sci 68:1553-1561

Kinghorn BP, Smith C and Dekkers JCM (1991) Potential genetic gains with gamete harvesting and in vitro fertilization in dairy cattle. J Dairy Sci 74:611-622

Nicholas FW and Smith C (1983) Increased rates of genetic change in dairy cattle by embryo transfer and splitting. Anim Prod 36:341-353

Chapter 22

Design of Crossbreeding Programs

Andrew Swan and Brian Kinghorn

Straight *versus* Crossbreeding

Much of this book is devoted to the improvement of animals through selection and mating within breeds. The other basic tool available for genetic gain is crossbreeding. While gains from selection and mating within breeds accumulate over generations, gains from crossbreeding occur immediately but are not cumulative. Consequently, crossbreeding provides flexibility in changing production and marketing environments. Selection and crossbreeding can be combined, and adoption of such a strategy gives breeders a much wider range of genetic material than within-breed selection, although crossbreeding programs are operationally more complicated. There is a diverse range of breeds and strains within livestock species suited to a variety of purposes. They are a valuable resource, and should not be ignored in livestock improvement programs.

The Benefits Of Crossbreeding

The benefits of crossbreeding may be split into a number of components:

The average breed effects of crossbreds. This component is very frequently of negative value, where one parental breed is superior for the production and marketing system of interest. However, the ideal genotype can be intermediate between existing breeds. For example, in dairy cattle a genotype with intermediate body size may best suit the economic environment. In meat production, certain markets may require carcases intermediate in size of cut and fat cover to existing breeds. The averaging of breed effects can also be important where two traits are negatively correlated across breeds, and where they act multiplicatively in the expression of economic value. This is the case in *Bos taurus* (high producing) x *Bos indicus* (stress and disease resistant) crosses for beef production.

Heterosis. Crossbred populations exhibit heterosis for a range of traits. That is, they perform better than the average of their parent breeds. The percentage increase in performance differs markedly between traits, species, and the breeds involved. Heterosis for production traits is usually in the range -5% to 10%, whereas heterosis for traits related to fertility is usually in the range +5% to 25%. The effect of heterosis on the total production system can be even more than this, as effects accumulate over traits through both direct and maternal expressions. For a composite trait such as weight of calf weaned per cow joined, it has been estimated (Gregory and Cundiff, 1980) that crosses among *Bos taurus* cattle show 23%

heterosis, while *Bos taurus* x *Bos indicus* crosses can show up to 50% heterosis. Maternal heterosis accounts for up to 60% of the total gain in these crosses.

The complementation of sire and dam effects for meat production. The proportion of total costs required for maintaining breeding females can be very high, especially where fecundity is low. For a given size of female, larger sires and therefore larger and faster growing slaughter progeny mean a lower dam cost of meat production, all other traits being equal. In this context, a crossbreeding system aims to use breeding males which are large, but not too large for dystocia in mates to be a problem.

The adoption of a crossbreeding strategy provides the only opportunity to make full use of all available genetic resources. The range of genetic material available is greater when more than one breed is involved. This can give greater flexibility in identifying and developing an efficient breeding program. Selection intensities can be higher, and inbreeding problems can be largely avoided. However, in some production systems and species there may be a single breed of notably superior productivity, such that no other breed can contribute to higher yields. This is most likely to be of importance where other breeds cannot even contribute for traits such as disease resistance and adaptability. Further, planned crossbreeding programs involve more complicated logistics than straightbreeding programs simply because there are now two or more gene pools involved. Often this issue is not sufficiently recognised.

Problems Associated With Crossbreeding

Although there are considerable benefits from crossbreeding in most species and production systems, there are logistical and biological factors which can inhibit the uptake of crossbreeding:

Extra management. Crossbreeding programs within a single production unit can become complicated because of the need to maintain a number of straightbred and crossbred types in separate management groups. Although there are different crossing systems suited to different levels of management, generally more intensive management is required in crossbreeding programs than in straightbreeding programs.

Importation of breeding stock. Where crossbred seedstock must be imported, extra costs are incurred compared to self-replacing straightbred herds or flocks. Also, information on the genetic merit of crossbred seedstock is not currently available (see Chapter 12).

Reproductive rate. Reproductive rate of the species involved can inhibit the use of certain types of crossing systems. For crosses involving species with low reproductive rates such as sheep and cattle, the proportion of animals derived from purebred matings can be high, reducing the number of saleable animals of the desired crossbred type. This is not such a problem in species with high reproductive rates such as pigs.

Crossbreeding programs should not be initiated unless economic returns are higher than from straightbreeding programs. However, with careful consideration of benefits and costs, appropriate crossbreeding programs can be designed for many production systems.

Crossbreeding Systems

The three basic crossbreeding systems are:

- Specific crossing systems
- Rotational crossbreeding
- Formation of composite or synthetic breeds

Crossbreeding is also used to expand exotic breeds by backcrossing, and to introduce new genetic material to an existing breed. The choice of crossbreeding system is governed by several factors:

- Reproduction rate of the species
- The importance of heterosis and breed effects
- Economic factors, including the cost of more intensive management

Specific crosses. In specific crossing systems, straightbred parent lines must be maintained to generate the desired crossbred animals each mating. Specific crosses include F_1 (or first) crosses, backcrosses, 3-way crosses, 4-way crosses etc.

Under a dominance model of heterosis (see Chapter 12), F_1 crosses express direct heterosis fully, while 3-way crosses using F_1 dams, and sires of a third breed, express both direct and maternal heterosis fully. A 4-way cross using F_1 sires and dams may be used to take advantage of heterosis from both the sire and dam lines. By choosing sire and dam breeds appropriately, breed effects and complementarity give additional benefits: for example, the sire breeds may be chosen for their meat producing ability, while the dam breeds may be chosen for maternal ability and low maintenance costs. However, because straightbred matings are continually required to produce parents, especially females, specific crossing systems are better suited to species with high fecundity, as a smaller proportion of the population is required to maintain the straightbred parental lines. However, in some situations, it may be economically viable to produce F_1 females in species with low fecundity. One such example is the production of Border Leicester x Merino ewes for the Australian meat sheep industry using aged Merino ewes from the wool industry.

Backcrossing is mainly used for grading up to new breeds. This system involves mating sires of the replacement breed to dams of the existing breeds. The resulting F_1 dams are then mated to sires of the replacement breed, and so on. This system is useful in introducing novel breeds into a country. Cattle breeds such as the Simmental and the Charolais have been introduced into Australia in this way. New genetic material can also be introduced into existing breeds by backcrossing. In this case, once the introduction of genetic material is completed, the progeny are backcrossed to the existing breed.

Rotational crossing. Rotational crossing, also called criss-crossing, is a form of backcrossing in which the sire breed is alternated in a specific sequence. For example, a 2-breed rotation between breeds A and B may proceed by making an F_1 cross with A as the sire breed. In the next generation a backcross is made using B as the sire breed, and in the third generation A is used again. At equilibrium, the breed composition in a 2-breed rotation stabilises at 2/3 of the sire breed and 1/3 of the maternal grandsire breed. The heterosis (direct and maternal) expressed in a 2-breed rotation under a dominance model is expected to be 2/3 of the heterosis expressed in the F_1 cross between the two breeds.

The advantage of rotational crossing over specific crossing systems is that only sires are required from purebred matings: crossbred dams are self-replacing. The main disadvantage is the large variation in breed composition between generations. Thus for marketing and management purposes, similar breeds may have to be used. This reduces the opportunity to take advantage of breed effects and complementarity. A modification to overcome this problem is to use a terminal sire on the older females in the herd. Older females are less susceptible to dystocia that may arise with certain terminal sire breeds. A further disadvantage of rotations is the large number of crossbred types formed before the system reaches equilibrium.

A variation of rotational systems is the periodic rotation (Bennet 1987a). Whereas conventional rotations use sire breeds equally, periodic rotations use sire breeds unequally. In other words, in one complete cycle of an n-breed conventional rotation, there will be n generations with each sire breed used once, while in periodic rotations, there can be more than n generations because each sire breed can be used more than once. For example, in one cycle of a 3-breed conventional rotation the sequence of sire breeds used is A,B,C, while one cycle of a periodic rotation may use the sequence A,B,B,A,A,C. This means that the possible sequences of n-breed periodic rotations are numerous. Periodic rotations can be found which show less variation in merit between generations than conventional rotations. Although periodic rotations generally express less heterosis, they can exploit breed effects, for example, by using poorer breeds less. As a result, under certain conditions, where breed effects are larger than heterosis, periodic rotations can be found that exceed conventional rotations in merit.

Composite breeds. The third basic type of crossbreeding system involves formation of composite breeds. This has been advocated for small herds and flocks because the management problems of specific crosses and rotations can be avoided: these systems have several different mating types with respect to breed and these must be managed separately, while composites can be managed as straightbred once the initial cycles of crossing are complete. Composites express less heterosis than specific and rotational crossbreeding systems using the same breeds. However, heterosis expressed can be increased by using a terminal sire breed over the older composite females. A further alternative is the sire-breed rotation (Bennet 1987b), where the sire breeds are rotated in some sequence, and the breed pedigree of the dams is ignored. Appropriate choice of sequence and of culling age of females can result in a system with more heterosis than a similar composite.

For a given set of breeds there will be an optimal composite. The proportions of each breed in the optimal composite are determined by balancing the value of increased heterosis against inclusion of more breeds which may be inferior. The optimal composite can be calculated using information on breed effects and heterosis. Alternatively, under appropriate conditions, a mating strategy accounting for both within- and between-breed effects establishes a nearly optimal composite. This approach is discussed in the next section.

One potential benefit of composites is that increased genetic variance may occur, leading to greater selection response. Disadvantages are that they may be more susceptible to the breakdown of favourable combinations of genes in the purebreds (a phenomenon known as epistatic loss) than systems which use purebred sires, and inbreeding may offset the heterosis regained, particularly in small populations.

A Dynamic Approach To Establishing Crossbreeding Programs

A crossbreeding program can be established without actually aiming at a specific crossbreeding system (Kinghorn 1986). This can be achieved using a mating strategy where the expected progeny merit is calculated for all possible matings among candidate animals. Expected progeny merit can be calculated given information on breed effects and heterosis, and estimated breeding values (EBVs) for all animals. Across-breed genetic evaluations may be used to generate this information (see Chapter 12). A cost correction must also be included. For example, the cost of importing animals should be accounted for.

Using this method, the optimal crossbreeding system will be established for the prevailing genetic and cost parameters, as demonstrated in Table 22.1. For example, if direct and maternal heterosis are high and importation costs are low, a 3-way specific cross will be established importing F_1 females, and males of a terminal sire breed. If the cost of importing F_1 females is prohibitive, a rotational crossbreeding system may be established. If there is significant within breed variation, a composite may be established. The proportions of each breed in such a composite will approach optimal levels.

Table 22.1 Optimal crossbreeding systems under different genetic and cost conditions, management costs or costs of importing crossbred seedstock relative to maintaining self-replacing lines

Direct heterosis	Maternal heterosis	Within-breed genetic variation	Cost	System chosen
Low	Low	High	High	Straightbreeding
High	Low	Low	Low	F_1 cross
High	High	Low	Low	3-way cross
High	High	Low	High	Rotation
High	High	High	High	Composite

As with straightbred seedstock production, such dynamic crossbreeding programs require an efficient performance recording system, including individual animal identification and careful structuring of management groups.

Practical Application of Crossbreeding in Livestock Industries

The dynamic approach described above provides a useful tool for the design of crossbreeding programs. However, such a service is not provided by genetic evaluation systems currently available, and may not be for some years. More research must be done on across-breed genetic evaluation, and on the economic and logistical factors involved before this is possible. In this section, the crossbreeding systems most suited to the major livestock industries are discussed.

Beef cattle. Beef cattle are not ideally suited to specific crossing systems. Their low reproductive rate means that a large proportion of the herd must be kept as straightbreds to generate the desired crossbred progeny. Also, the relatively low level of management under range conditions and small herd size do not lend themselves readily to crossbreeding.

Crossbreeding in more stressful environments such as tropical regions has well recognised benefits. *Bos taurus* cattle are generally poorly adapted to tropical environments but have a high growth potential in stress-free conditions, while the reverse is observed in *Bos indicus* cattle. Crossing the two types produces cattle which are both well adapted and have a high growth potential. Suitable crossing systems for these environments are determined primarily by the extensive management systems used. Specific crossbreeding systems are not practical because of the need to manage several different crossbred types. These systems may also be at a disadvantage if straightbred lines must be maintained, especially *Bos taurus* breeds. Traditional rotations do not have this problem, but it is still necessary to manage several different crossbred types separately. Perhaps the best possibility is the formation of composite *Bos taurus* x *Bos indicus* breeds. These composites are already popular in some countries, and there are several recognised breeds which have been formed in this way, e.g. Brangus, Braford, Simbrah, Charbray, and Belmont Red. The use of sire-breed rotations is another practical alternative, as all that is required is to purchase sires of different breeds in certain years. Some producers already use this system, although not in a formalised manner.

Crossbreeding can also provide significant gains in temperate beef industries. The benefits of *Bos taurus* x *Bos indicus* crosses are not as marked, but crosses among *Bos taurus* breeds can show significant amounts of heterosis, and provide a flexible way of meeting requirements in a dynamic market. Specific crossing systems are unsuited to a single production unit, because of relatively small average herd size and the low reproductive rate of cattle. These two factors result in very few saleable cattle of the desired terminal cross, and little opportunity for selection in the straightbred parental lines. Rotational systems are more viable because all reproducing females are crossbred, and more of the animals sold are of the desired crossbred type. A terminal sire breed can be mated to a portion of the herd for further gains. However, rotational systems still require extra management compared to straightbreeding.

A viable option for temperate beef cattle industries may be to produce different crosses in different geographical regions, with the appropriate cross being determined by factors such as climate, pasture productivity and management systems. For example, producers in less productive regions could concentrate on breeding F_1 females which are then marketed to producers in more favourable regions. These females would then be joined to terminal sires of a breed appropriate for the targeted market. The across breed genetic evaluation methods described in Chapter 12 would be useful for choosing these terminal sires. A recent trend in the Australian beef industry has been the popularity of F_1 female sales, where this type of structure is gaining momentum.

Dairy Cattle. Crossbreeding for milk production has little value in dairy industries located in temperate regions. As the Holstein-Friesian breed produces higher milk volumes and absolute yields of fat and protein than any other breed or cross, there is a trend to use it exclusively in temperate production systems. However, the dairy industry is well placed to use a dynamic approach to crossbreeding as described in the previous section. The extensive use of AI makes a range of sires from different breeds readily available to individual producers. AI has also allowed large across-herd genetic evaluations, which could be extended to across-breed evaluations. This would give producers the option of directly comparing animals of different breeds and crosses. Consequently, dairy producers could adopt a strategy of importing superior animals without aiming at a particular crossbreeding system. This may lead to the formation of composite breeds, to take advantage of both within breed variation and crossbreeding effects. This has commenced in Australia with the recent introduction of a national across-breed genetic evaluation for the red breeds.

One potential use for crossbreeding in dairy cattle is to produce animals for the beef industry. Dairy producers can mate a portion of their herd to bulls from beef breeds. The male progeny show high rates of gain in feedlots, while the female progeny are excellent F_1 dams in specific crossing systems. This would be complementary to the system proposed above for the beef industry, where F_1 females are marketed in special sales. The main difficulty is in rearing the animals to a marketable age, which can be expensive.

In environments such as the tropics, crossbreeding for milk production does have benefits, especially where milking cows are subject to stress. For example, high producing Friesians may have re-breeding problems and may be more susceptible to disease than lower producing native breeds. Many dairy cattle are subject to these conditions in developing countries, where there is little use of herd recording, often very small herd size, and limited use of intensive management procedures. In these production systems, composites or native breeds are favoured over specific crossing systems.

Meat Sheep. Crossbreeding in meat sheep industries is used extensively in some countries. For example, a 3-way crossing system is used almost exclusively in Australia, where aged Merino ewes from the wool industry are mated to Border Leicester rams to produce F_1 ewes, which are then mated to rams of a terminal sire breed such as the Dorset. The main limitation to the system is the quality of the sire breeds, the terminal sire breeds, and the Border Leicester. The terminal sire breeds are generally divided into small flocks, leading to high rates of inbreeding, and limiting the genetic gain which can be made because of reduced selection intensity. However, the breeders of these sheep are becoming more aware of these problems through the introduction of scientific breeding principles via the industry's genetic evaluation program, LAMBPLAN.

An interesting development in Australia is the introduction of the Booroola gene for prolificacy into the Border Leicester, a project undertaken by CSIRO. Ewes carrying two copies of the gene (homozygotes) are highly prolific, while ewes carrying one copy (heterozygotes) are intermediate. Heterozygotes are therefore easier to manage in a typical meat sheep production system. The system proposed is to mate Border Leicester rams homozygous for the Booroola gene to Merino ewes. The resulting F_1 ewes will have a single copy of the gene, and therefore a manageable increase in prolificacy.

Wool sheep. The Merino breed is superior to all other breeds and crosses for the production of fine apparel wool, so crossbreeding is of little or no interest to wool growers. The possible exception is in enterprises where both wool and meat production are considered important. Several dual-purpose composite breeds have been developed for such enterprises, including the Corriedale and the Polwarth in Australia, and the Coopworth in New Zealand. While such breeds are used widely in New Zealand, they have not found widespread acceptance in Australia due to the economic superiority of the Merino for wool production and of 3-way crossing system for meat production, as described above.

The Australian Merino industry has been characterised by the development of various strains and bloodlines within strains, which have formed in geographical isolation. To date there has been little experimental work involving crossbreeding between strains or bloodlines. The limited number of studies show heterosis does occur in crosses between strains, most significantly for reproduction traits. However, further studies are needed for more conclusive evidence. The most appropriate crossbreeding system may be the development of composites, due to the extensive management systems used in the industry.

Pigs. Crossbreeding has well recognised advantages in the pig industry, and its use is widespread internationally. Specific crossing systems are favoured due to the high reproductive rate of pigs, only a small proportion of animals need be purebred, and the intensive management systems. 3-way crossing systems are most common, with boars from a terminal sire line mated to F_1 sows derived from maternal lines, utilising the considerable heterosis for reproduction. However, 4-way specific crosses are also used, as F_1 boars can show higher libido and conception rate than straightbreds. Another situation where F_1 boars may be of use is where a high producing terminal sire line is carrying the halothane gene, a recessive gene related to stress susceptibility and poor meat quality. This line may be crossed to a line which is free of the gene thus ensuring there are no homozygous carriers in the F_1 boars. Because the gene is recessive, the stress effects only occur in animals homozygous for the gene. Providing there are no carriers among the F_1 sows, there will be no homozygotes among the progeny.

Guide for Consultants

- Crossbreeding provides the opportunity to use a much wider range of genetic material than straightbreeding, and allows a more flexible approach in changing marketing and production environments.
- The value of crossbreeding is derived from the averaging of breed effects, heterosis, and complementation of sire and dam effects.
- There are a number of crossbreeding systems to choose from, including specific crossing systems, rotations, and the development of composites. Crossbreeding programs should be designed carefully, with consideration given to the importance of breed effects and heterosis, the fecundity of the species, the management systems which can be used, and economic and logistical factors.

References

Bennet GL (1987a) Periodic rotational crosses I. Breed and heterosis utilization. J Anim Sci 65:1471-1476

Bennet GL (1987b) Periodic rotational crosses III. Sire-breed rotations with overlapping generations among dams. J Anim Sci 65:1487-1494

Gregory KE and Cundiff LV (1980) Crossbreeding in beef cattle: evaluation of systems. J Anim Sci 51:1224-1242

Kinghorn BP (1986) Mating plans for selection across breeds. 3rd World Conf Genet Appl Livest Prod Vol XII 233-244

PART V: The Breeding Business

Chapter 23

Other Economic Considerations in Animal Breeding

Keith Hammond

Background

We commenced Chapter 1 listing the fundamental considerations to profitable livestock production and the avenues available to the livestock producer to maintain and improve profitability, viz. winning market share and increasing dollar return per dollar spent by utilising better equipment and processes, obtaining and utilising better information, and making genetic change. Animal Breeding concerns the manipulation of biological differences between animals over time using approaches aimed at maximising profitability in both the short term and the longer term. Some of these biological differences are genetic or inherited, whilst others are non-genetic or environmentally induced. Genetic differences may occur between breeds, between crosses, and between animals within breeds and within crosses. Environmentally induced differences may be expressed once only, e.g. the result of a change in feeding, or repeatedly, e.g. the lifetime impact of feeding on mammary gland development and resulting milk production and progeny growth following each subsequent parturition.

Hence, economic considerations in Animal Breeding involve the use of a range of reasonably standard evaluation procedures. They vary depending on whether the costings involved are directed at:

- The current herd.
- The impact of long-term investment in breeding.
- Cash flow considerations.
- Structural considerations of, e.g. within-herd or flock age structures.
- Seedstock producer *versus* buyer considerations.

Here, we briefly comment on these alternative costings without detailing the procedures.

Of course defining and establishing the sensitivity of the breeding objective is an economic rather than a genetic problem; as the primary aim is to establish the economic values for each input and output trait in the objective. This has already been dealt with at length in Chapters 13 to 18.

The Seedstock Producer

- Will need to examine the **benefits** and **costs** of :
  - Selling seedstock, including the time consuming establishment of a clientele.

- Registering some or all of the herd or flock with a breed society.
- Objectively *versus* subjectively measuring traits and being involved in performance recording operations.
- Subdivision fencing to enable single sire joining - do the benefits outweigh the additional fencing and management costs?
- Importing animals, semen, or embryos, - sometimes it is only the grass that is greener rather than the genes being better! How to reliably establish genetic superiority?
- What base animals to use in initiating a seedstock producing operation: The best from the best herds or flocks? The best semen on average females? Average or any sires from the best herds or flocks? How many herds or flocks to sample? Remember, seedstock production involves long-term investment - both speed of change and handicap at the start are important!
- The alternative breeding structures which could be used to achieve genetic change and so remain competitive. Alteration of structure will change one or more of: the accuracy of selection, intensity of selection, generation interval and level of inbreeding, on one or more of the parent to offspring pathways along which genes are transmitted between generations.
- The vast majority of benefit-cost analyses of alternative breeding structures (designs of breeding programs) only consider long term return on the investment in breeding. In the process these analyses also commonly examine the impact of additional variables such as the time horizon over which the benefit-cost analysis is done, the inflation free discount rates used to discount costs and returns back to present day comparable values, whether the demand for breeding stock will reduce (the size of the industry is decreasing), remain stable or increase over the time horizon being considered, the value of the population at the end of the costing, and so on.
- However, the seedstock producer or artificial breeding organisation also needs to live from year to year! Hence the ability of alternative breeding structures to achieve continuity of cash flow is also extremely important. As an example, the artificial breeding organisation is not so concerned about overall rate of genetic progress but whether the organisation has a sire at the top of the current and next sire summary list, and the semen producing capacity of these sires.
- Cash flow considerations are more important in the larger animal species with longer and overlapping generations, where there is less flexibility to move quickly and make change to the breeding operations, to again become competitive.
- The combined use of Artificial Insemination and an effective public relations program still offer the best means for moving into a seedstock production and marketing operation in most industries.

The Commercial Producer

- Benefit-cost analyses of straight-breeding *versus* cross-breeding and the design of these breeding operations will need to consider :
  - Current and future market demand for end products, and the access to and stability of these markets.
  - The availability of seedstock for producing replacement straight-bred or cross-bred females and replacement males, and the longer term reliability of these supplies.
  - Whether to breed replacement sires, buy all replacements, or buy just one top sire or semen each year and use this to breed and use one or more crops of sons to produce commercial product.
- Establish the value of a unit of EBV for each trait for all measurements of importance, i.e. the index, for the current market and for new markets.

The Breeding has been Done For Today!

Remember even in purchasing seedstock for commercial production, the results of these buying decisions will be reaped over one or more crops of offspring in the future, and with cattle this will be several years ahead!

For the buyer of seedstock, the modern genetic evaluation operations in industries such as the dairy and beef industries, which evaluate genetic merit across herds, operate by reducing the area of non-discrimination i.e. by increasing the buyers ability to discriminate between herds, flocks or AI studs on genetic merit.

For the seedstock producer the breeding decisions have already occurred for today and those currently being made relate to the next two to three generations (6 - 12 years in cattle!). Nor can the seedstock producer circumvent this inherent problem of Animal Breeding; whether or not formal calculations and designs are implemented, breeding is concerned with future markets and effective marketing of seedstock.

These calculations and decisions will be predictions and as such involve risk. This is discussed in Chapter 24.

Chapter 24

Management of Risk

Markus Schneeberger

Classically Animal Breeding programs have dealt with maximising expected genetic gain and have generally neglected risk. But risk is involved in animal breeding at various levels, as pointed out by Anderson (1988) and Tier (1988). Prices for both inputs and outputs, climatic factors or diseases are rather unpredictable. Risk also arises from genetic sources: even if the breeding value of an animal was known perfectly it cannot be predicted what half of the animal's genes will be transmitted to a particular offspring. Predicting the response to selection for the breeding program as a whole includes the combination of risk associated with all individual breeding events in the program.

The Risk About Selection of Replacements

Consider the risk associated with genetic evaluation. The measure for this risk is the accuracy of a predicted breeding value. The higher the accuracy, the lower is the associated risk. Some preliminary comments on the use of accuracies or reliabilities of EBVs or EPDs were included in Chapter 4. Here, we combine this information using a utility function. The attitude towards risk differs from one person to another. It can be expressed by a utility function of the form

$$U = E + b V,$$

where

- U = utility
- E = expected income
- V = risk or variance of the income
- b = the weighting for risk relative to expected income.

A person with $b = 0$ is risk neutral, $b<0$ indicates risk aversion, and $b>0$ would indicate a risk prone attitude.

    Methods to assess risk preferences were summarised by Cochran *et al.* (1990), e.g. they can be assessed by direct elicitation: the decision maker is presented choices between alternative plans with varying expected incomes and risks (e.g. Schneeberger and Freeman 1980). Utility functions can also be derived from observed economic behaviour. An example for this is given by

Schneeberger et al. (1981) who derived a utility function for buyers of dairy semen from actual semen sales and their associated expected incomes and risks.

When selecting sires to mate a herd of cows, risk can be reduced by selecting sires with higher accuracies and by selecting a larger number of sires, thus by diversifying. This is equivalent to selecting a portfolio of stocks on the stock market where each stock has an associated expected income and risk expressed as the variance of income. Portfolio theory (Sharpe 1970) deals with this.

The problem is to maximise utility,

$$U = p'x + k\ x'Q\ x, \tag{1}$$

where

U = utility,
p = a vector with expected incomes for each activity or sire,
x = the vector of solutions for each activity or sire,
Q = the matrix of variances and covariances among the activities or sires, and
k = the weight for variance relative to expected income.

Equation (1) is maximised subject to a set of constraints,

$$A\ x \leq b,\ \text{and}\ \ x \geq 0,$$

where A is a matrix of constraints and
b is the vector of right-hand sides of the constraints.

A Portfolio of Sire Usage

This method is illustrated by the example given in Schneeberger et al. (1982). Four sires are available for selection with the following expected incomes p and variances Q, assuming zero covariances among them:

$$p = \begin{bmatrix} 84 \\ 95 \\ 140 \\ 61 \end{bmatrix} \qquad Q = \begin{bmatrix} 598 & 0 & 0 & 0 \\ 0 & 897 & 0 & 0 \\ 0 & 0 & 6132 & 0 \\ 0 & 0 & 0 & 3440 \end{bmatrix}$$

The relative weight for risk or variance of income was chosen as $k = -0.02$. The objective is to find the portfolio of sires (i.e. the proportions of cows that are to be mated to each of the four sires) that maximises utility,

$$U = \mathbf{p'x} - 0.02\, \mathbf{x'Q\, x}.$$

The ith element of the solution vector $\mathbf{x}$, x_i, is the proportion of cows mated to the ith sire. Constraints have to be introduced to ensure that not more than all cows can be mated ($\Sigma\, x_i \leq 1$), and each of the proportions x_i has to be ≥ 0. The problem, therefore, is to maximise

$$U = 84x_1 + 95x_2 + 140x_3 + 61x_4 - 12x_1^2 - 18x_2^2 - 123x_3^2 - 69x_4^2$$

subject to $x_1 + x_2 + x_3 + x_4 \leq 1$
$x_i \geq 0,\ i=1,2,3,4$

The solution is:

$x_1 = .25$
$x_2 = .50$
$x_3 = .25$
$x_4 = 0$

The expected income and the variance of the income are

$$\mathbf{p'x} = 104,\ \mathbf{x'Q\, x} = 645,$$

and the utility is

$$U = 104 - 0.02 \times 645 = 91.$$

Thus, utility for this specific value of k is maximised when a quarter of the cows are each mated to sire 1 and 3, and half the cows are mated to sire 2. Sire 4 is not used at all.

This result, of course, is only valid for one specific utility function, i.e. the function with $k = -0.02$. A more generally applicable result can be found by determining the set of portfolios that minimizes the risk for each possible value of expected income. This set is called the efficient set of portfolios and includes the optimum solution for any value of k in the utility function.

The problem can then be formulated as follows:

Minimise $V = x'Q\ x$

subject to $p'x = ß$
$\ \ \ \ \ \ \ \ \ \ \ \ \ \ \ A\ x \leq b$
$\ \ \ \ \ \ \ \ \ \ \ \ \ \ \ x \geq 0$

ß is varied according to the specific problem which in the above presented example is from 0 to 140, the maximum single value in **p**. It suffices to find the solutions for **x** for the **b**s where a new sire enters or leaves the efficient set of portfolios. These solutions are presented in Table 24.1. All solutions between these points can be obtained by linear interpolation. The optimum solution is in the neighbourhood of U=89, and by linear interpolation, it can be derived that it is in fact the solution for U = 91 as obtained above.

Table 24.1   Efficient sets of portfolios for values of the expected income at which a sire enters or leaves the set.

x_1	x_2	x_3	x_4	ß	V	U
0	0	0	0	0	0	0
0.49	0.37	0.08	0.06	91	318	85
0.42	0.43	0.15	0	97	404	89
0	0.59	0.41	0	114	1350	87
0	0	1.00	0	140	6132	17

The trade-off between expected income (ß) and risk or variance of income (V) becomes apparent in Table 24.1. The highest expected income is realised when only the best sire, sire 3 is used, but risk is also highest. When risk is reduced, inferior sires have to be added to the portfolio. The points in Table 24.1 can also be graphically represented (Figure 24.1) as the E-V (expected income vs. variance) frontier. Expected income, $p'x$, is found on the abscissa, and standard deviation (SD) of income, $(x'Q\ x)^{0.5}$, on the ordinate. The optimum solution is determined by plotting iso-utility curves and finding the point of tangency with the E-V frontier. Iso-utility curves for the utility function with k = -0.02 are plotted in Figure 24.1 for U=90, U=100 and U=110, and the optimum solution (the point of tangency) is found at expected income = 104, SD of income = 25, and U = 91.

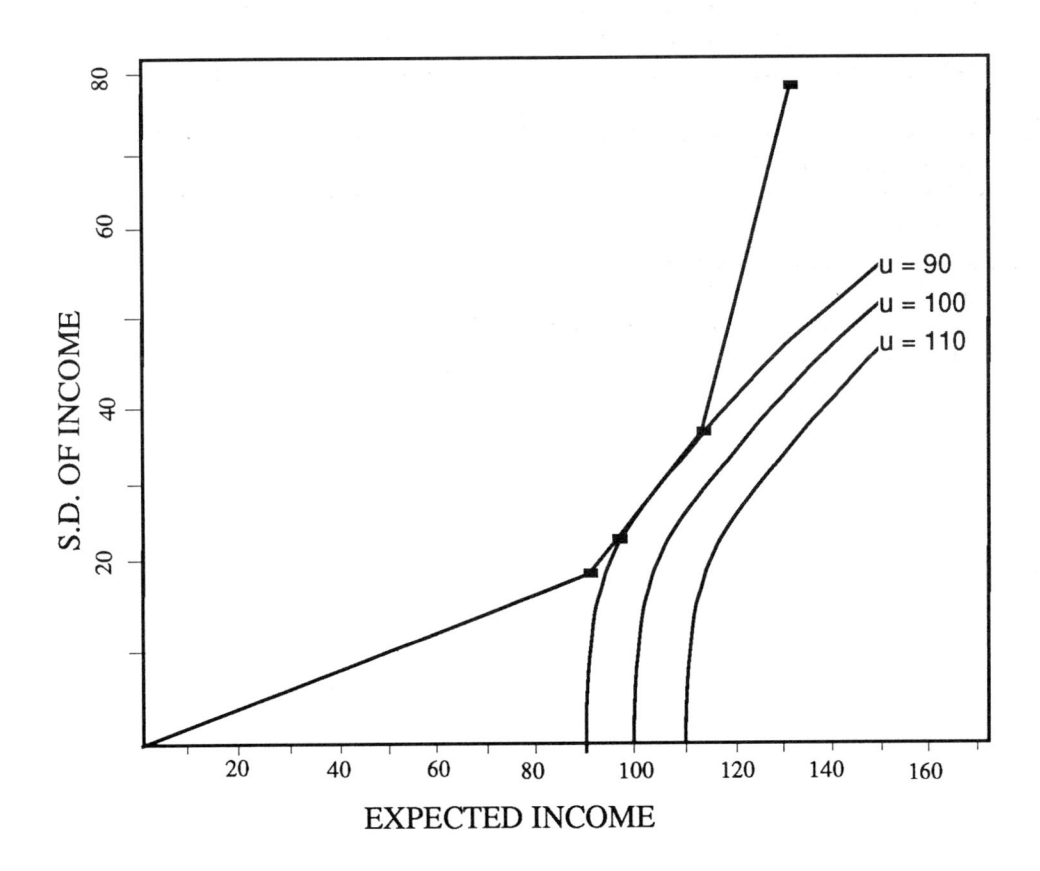

Fig. 24.1 Expected income versus variance frontier with points at which a sire enters or leaves the solution and iso-utility (U) curves for the example with 4 sires of Schneeberger *et al.* (1982)

This method has been applied by Schneeberger *et al.* (1981) to data of 285 Holstein sires. More recently, Everett (1989) developed a practical application where different portfolios of dairy sires representing different levels of risk are presented to buyers of semen of dairy bulls.

References

Anderson JR (1988) Accounting for risk in livestock improvement programs. Proc 7th Conf AAABG, Armidale, 26-29 Sept: pp 32-41

Cochran MJ, Zimmel P, Goh SC, Stone ND, Toman TW, Helms GL (1990) An expert system to elicit risk preferences: The futility of utility revisited. Computers and Electronics in Agriculture 4: 361-375

Everett RW (1989) Breeding dairy cows is like playing the stock market. Hoard's Dairyman, March 25: 248, 279

Schneeberger M, Freeman AE (1980) Application of utility functions to results of a crossbreeding experiment. J Anim Sci 50: 821-827

Schneeberger M, Freeman AE, Boehlje MD (1981) Estimation of a utility function from semen purchases from Holstein sires. J Dairy Sci 64: 1713-1718

Schneeberger M, Freeman V, Boehlje, MD (1982) Application of portfolio theory to dairy sire selection. J. Dairy Sci 65: 404-409

Sharpe WF (1970) Portfolio theory and capital markets. McGraw-Hill, New York

Tier B (1988) Accounting for risk in livestock improvement programs. AGBU. Memo

Index

$EBV, 109
$INDEX, 17, 109, 186

ABV Australian Breeding Values
ABV Australian Breeding Value
ADG Average Daily Gain
ADHIS Australian Dairy Herd Improvement Scheme
AGBU Animal Genetic Breeding Unit
AI & Across-Herd or Flock Evaluation, 72
AI & MOET in Open Nucleus Schemes, 222
AI Artificial Insemination
AI Maximum Improvement, 217
AI in Males & Selection Intensity, 219
Accounting for Competitive Position, 136
Accounting for Feed Costs, 136
Accuracy Estimated Breeding Values, 55, 64
Accuracy Selection Program, 207
Across Flock Evaluation, 87
Across Herd or Flock Comparisons, 71
Across- Versus Within-Breed Evaluation, 111
Across-Breed Evaluation Industry Application, 117
Across-Breed Genetic Evaluation, 111
Across-Flock Genetic Evaluations, 92
Across-Herd or Flock Analyses, 75
Across-Herd or Flock Analyses for Meat Yield & Quality, 76
Across-Herd or Flock Operations, 73
Across-Herd or Flock Procedures Genetic Evaluation, 72
Age Limits BREEDPLAN, 81
Age at Puberty Heritability, 49, 104
Age at Sexual Maturity Heritability, 49
Allele Introduction, 215
Allelic Contribution Heterosis Expression, 113
Alternate Evaluation Procedures, 57
Analyses Across-Herd or Flock, 75
Analysis Procedures, 10
Animal Breeding & Economics, 121
Animal Breeding Concepts, 23
Animal Breeding Decision Areas, 13
Animal Breeding Does it Work?, 03
Animal Breeding Economic Considerations, 237
Animal Breeding History, 05
Animal Breeding Information, 07
Animal Breeding Modern Relevant Developments, 04

Animal Breeding New Era, 09
Animal Breeding Technological Developments, 10
Animal Breeding What is it?, 01
Animal Breeding Which Genetics is Important, 19
Animal Genetics & Breeding Unit, 17, 77
Animal Management Performance Recording, 29
Animal Model BLUP, 61
Animal Model Maternal Effects, 66
Animal Model Multiple Trait, 66
Animal Model Single Trait, 66
Application Crossbreeding in Livestock Industries, 232
Applications B-OBJECT, 148
Approaches to Breeding Objectives in other Countries, 174
Approaches to Use Breeding Objectives, 155
Artificial Breeding & Performance Recording, 41
Artificial Insemination & Selection Intensity, 219
Artificial Insemination Technology, 11
Assumed Genetic Parameters, 147
Attitude Change Breed Association, 78
Australian Breeding Value, 17, 97
Australian Dairy Herd Improvement Scheme, 07, 100
Average Daily Gain, 185
Average of EBVs of Parents, 54

B-OBJECT, 17, 18
B-OBJECT Applications, 148
B-OBJECT Examples, 149
B-OBJECT Steps in Using, 152
B-OBJECT Uses, 148
B-OBJECT Beef Cattle, 141
B-OBJECT Elements, 141
BIF Accuracy Estimated Breeding Values, 64
BLUP Animal Model, 61, 65
BLUP Best Linear Unbiased Prediction,
BLUP Classification of Models, 60
BLUP Multiple-Trait Model, 61
BLUP Obtaining the Solutions, 63
BLUP Pigs, 108
BLUP Single-Trait Model, 61
BLUP Sire Maternal Grandsire Model, 60
BLUP Sire Model, 60
BLUP Statistical Method, 59

BREEDPLAN, 07, 18
BREEDPLAN 600, 79
BREEDPLAN 900, 80
BREEDPLAN Genetic Parameters, 79
BREEDPLAN History, 78
BREEDPLAN International, 58
Backcrossing, 229
Backfat Heritability, 104
Backfat Probe Heritability, 49
Bakewell Robert Animal Breeding, 04
Banks R Breeding Objectives in Meat Sheep, 169
Banks R Genetic Evaluation in Meat Sheep, 89
Barwick S Breeding Objectives for Beef Cattle, 141
Barwick S Introducing Economics to Modern Animal Breeding, 121
Base for Breeding Values, 96
Basic Breeding Biology, 02
Beef Breeding Objective and Selection Index Package, 141
Beef Cattle Across-Breed Evaluation, 117
Beef Cattle Australia Calving Ease, 84
Beef Cattle Breeding Objectives, 141
Beef Cattle Crossbreeding, 232
Beef Cattle Genetic Evaluation, 77
Beef Cattle Heritability Estimates, 49
Benefit Cost Commercial Production, 239
Benefit Cost Seedstock Production, 237
Best Linear Unbiased Prediction, 06, 58, 108
Between Animals Genetic Differences, 02
Between Breeds Genetic Differences, 02
Between Crosses Genetic Differences, 02
Biological Differences, 02
Biological Indexes, 131
Biological Traits Influencing Returns & Costs, 134, 144
Biology Breeding Basic, 02
Birth Weight Heritability, 49
Birth Weight Pigs Heritability, 104
Body Weight Heritability , 49
Body Weight Economic Weight Dairy Cows, 179
Booroola Gene Crossbreeding, 234
Breed Associations Changing Attitude, 78
Breed Societies & B-OBJECT, 148
Breed Type Contribution to the Lamb Industry, 170
Breed or Buy, 15
Breeding Biology Basic, 02
Breeding Objective, 17, 121
Breeding Objective Data Sample Form, 161
Breeding Objective Establishment, 15

Breeding Objective Formulation, 142
Breeding Objective Pigs Traits Contributing, 184
Breeding Objective Steps & Defining, 133
Breeding Objective Trait, 122
Breeding Objectives & Economic Values, 162
Breeding Objectives & LAMBPLAN, 171
Breeding Objectives Beef Cattle, 141
Breeding Objectives Desired Gains, 155
Breeding Objectives Pigs Defining the Goal, 183
Breeding Objectives for Dairy Cattle, 177
Breeding Objectives in Meat Sheep, 169
Breeding Objectives in Wool Sheep, 155
Breeding Objectives in other Countries, 174
Breeding Objectives in the Pig Industry, 183
Breeding Operation Decision Areas, 13
Breeding Operations Information, 13
Breeding Operations Primary Components, 16
Breeding Program Design, 18
Breeding Program Effectiveness, 16
Breeding Program Single Genes, 213
Breeding Program Single Gene Traits, 214
Breeding Programs Information, 08
Breeding Situations First Categorisation, 143
Breeding Stock Importation & Crossbreeding, 228
Breeding System Identification, 133, 142
Breeding Value Base, 96
Breeding Value Reliability, 97
Breeding Value Sheep, 86
Breeding Values, 47
Breeding Values Overseas Bulls, 97
Breedplan Beef Cattle, 141
Budgeting Methods, 135
Bull Breeders & B-OBJECT, 148
Bull Buyers & B-OBJECT, 148
Buy or Breed, 15

Calculations of Selection Intensity, 199
Calving Ease & Milk Production, 99
Calving Interval Heritability, 49
Carcase Fat Thickness Heritability, 49
Carcase Quality Grade Heritability, 49
Carcase Traits & EBVs, 83
Carcass Fat Thickness Heritability, 49
Carrick M Breeding Objectives in Wool Sheep, 155
Carrick M Genetic Evaluation in Wool Sheep, 85
Cash Flow & Breeding Programs, 16
Categorical Traits Genetic Evaluation, 67

Cattle Beef Genetic Evaluation, 77
Cattle Dairy Breeding Objectives, 177
Central & On-Farm Tests Advantages
 & Disadvantages, 88
Central Agency Analysis, 75
Central Test Across Flock, 87
Central Testing Genetic Evaluation, 73
Central Testing Pigs, 105
Changing Attitude of Breed Associations, 78
Cloning Improvements Due to, 221
Cloning Maximum Improvement, 217
Close Herd Selection Program, 205
Commercial Producer Animal Breeding, 239
Communications Engineering, 10
Comparative Rankings of Young Bulls, 151
Comparison Across Age-Groups, 96
Comparison Across Breeds, 96
Comparison Across Herds, 96
Competitive Position Accounting for, 136
Complementation of Sire & Dam Effects, 228
Complex Evaluation Procedures, 115
Components of Across-Breed Genetic
 Evaluation, 116
Components of Calving Records, 81
Composite Breeds, 229, 230
Compound Traits Selection for, 130
Concepts in Animal Breeding, 23
Conclusion Breeding Objectives
 Meat Sheep, 175
Confirmation Pigs, 106
Constrained Indices, 137
Consultants Guide Across-Breed Genetic
 Evaluation, 119
Consultants Guide Breeding Objectives
 for Beef Cattle, 152
Consultants Guide Breeding Objective
 Pigs, 191
Consultants Guide Crossbreeding
 Programs, 235
Consultants Guide Economics and
 Animal Breeding, 138
Consultants Guide Genetic Evaluation
 Pig, 110
Consumer Requirements & Pricing
 Systems, 11
Contaminants & Pricing Systems, 11
Cost & Benefit Evaluation Selection
 Programs, 211
Cost Benefit Seedstock Production, 237
Cow Selection & EBVs, 99
Cross Versus Straight Breeding Programs, 227
Crossbreds Average Breed Effects, 227
Crossbreeding & Dairy Cattle, 233
Crossbreeding & Stressful Environments, 232

Crossbreeding & Temperate Beef
 Industries, 232
Crossbreeding Benefits, 227
Crossbreeding Effects, 111
Crossbreeding Problems, 228
Crossbreeding Program Design, 227
Crossbreeding Programs Dynamic
 Approach, 231
Crossbreeding Systems, 229
Crossbreeding for Meat Production, 228
Crossbreeding in Livestock Industries, 232
Crossbreeding in Meat Sheep, 233
Crossbreeding or Straight Breeding, 15
Crossing Systems, 229
Culling & Selection, 121
Culling Decisions, 15
Cutting Ability Heritability, 49

DNA Structure, 06
Dairy Cattle Across-Breed Evaluation, 118
Dairy Cattle Breeding Objectives, 177
Dairy Cattle Crossbreeding, 233
Dairy Cattle Heritability Estimates, 49
Dairy Industry Genetic Evaluation, 95
Data Combining On-Farm with Central
 Test Data, 75
Data Input Performance Recording, 32
Data Interpretation & Breeding Operation, 13
Data Management Performance Recording, 29
Data Processing & Breeding Operation, 13
Daughter-Dam Comparisons, 57
Decision Areas Breeding Operation, 13
Decision Areas in Animal Breeding, 13, 14
Decisions Animal Breeding Programs, 15
Decisions Performance Recording, 27, 29
Decisions Support Systems, 27
Definition Economic Value, 131
Definition Heritability, 195
Design Breeding Program, 18
Design Selection Programs, 210
Design of Crossbreeding Programs, 227
Design of Straight-Breeding Programs, 205
Designing Performance Recording
 Operations, 27
Desired Gains Breeding Objectives, 155
Desired Gains Index, 137
Diffusion Coefficients, 137
Direct Measurement Traits, 02
Direct Versus Indirect Measures, 35
Direction & the Breeding Business Pigs, 183
Direction of Genetic Change, 155
Discounted Gene Flow, 137
Discounting, 137

Distribution of Genetic Differences, 69
Dollar EBV Distributions, 218
Dollar EBVs, 52
Dollar Values Data Needed Dollar
   Value calculation, 160
Dominance Theory, 112
Dos & Don'ts in Performance Recording, 43
Dynamic Approach to Establishing
   Crossbreeding Programs, 231

EBV & Heritability, 47
EBV Calculation, 48
EBV Direct, 52
EBV Estimated Breeding Value
EBV Interpretation, 98
EBV Maternal, 52
EBV Reliability, 97
EBVd EBV Direct
EBVm EBV Maternal
EBVs the Future, 84
EBVs Accuracy, 55
EBVs Calculation for Milk Production
   Traits, 95
EBVs Scrotal Size, 82
EBVs Use, 98
EPD & Heritability, 47
EPD Estimated Progeny Differences
ET Embryo Transfer
Economic & Production Inputs to
   $INDEX, 189
Economic Considerations Animal
   Breeding, 237
Economic Efficiency, 135
Economic Factors Crossbreeding, 229
Economic Inputs to $INDEX, 186
Economic Value & Genetics, 157
Economic Value Assessment, 135
Economic Value Definition, 131
Economic Value Derivation, 134
Economic Value Estimates for Traits, 149
Economic Values & Breeding Objectives, 162
Economic Values Non-linearity, 136
Economic Values Relative, 132
Economic Values Traits Derivation, 134
Economic Values the Measurement Basis, 132
Economic Weights Breeding Objectives
   Dairy Cattle, 177
Economic Weights for Breeding
   Objective, 179
Economics & Breeding Objective
   Formulation, 131
Economics & Modern Animal Breeding, 121
Economics Whose to Consider, 135

Efficient Sets of Portfolios, 244
Egg Hatchability Heritability, 49
Egg Production Heritability, 49
Egg Weight Heritability, 49
Electronic Engineering, 10
Embryo Transfer Technology, 11
Engeler, 57
Engineering Electronic & Communications, 10
Epistasis, 19
Estimated Breeding Value Calculation
   an Example, 51
Estimated Breeding Values, 06, 17
Estimated Breeding Values Use, 50
Estimated Breeding Values Accuracy, 64
Estimated Breeding Values & $EBVs
   for Boars, 190
Estimated Progeny Differences, 17, 47
Estimation Methods, 135
Europe Western Breeding Objectives
   Lambs, 174
Evaluation Across Flock, 87
Evaluation Across- Versus Within-Breed, 111
Evaluation Methods Pigs, 106
Evaluation Procedures Historical
   Development, 57
Evaluation Structure in LAMBPLAN, 89
Evaluation Within Flock, 85
Evaluations Genetic Across-Flock, 92
Ewe Migration Pattern, 202, 203
Example Pig Production Unit, 188
Examples Selection Index, 126
Excitability Heritability, 49
Expected Income Versus Variance
   Frontier, 245
Experimental Design & Performance,
   Recording, 29
Extension Problems EBVs, 100
Eye Cancer Heritability, 49
Eye Muscle Area & LAMBPLAN, 90
Eye Muscle Genetic Evaluation, 172

FD Fibre Diameter
Face Covering Heritability, 49
Fat Depth & LAMBPLAN, 90
Fat Economic Weight Dairy Cows, 179
Fat Production Heritability, 49
Feed Costs Accounting for, 136
Feed Efficiency Pigs Heritability, 104
Female Parents Selection Differentials, 196
Fibre Diameter & LAMBPLAN, 90
Fibre Diameter Heritability, 49
Fibre Diameter in $EBVs, 52
Figure 1.1 History of Animal Breeding, 05

Figure 1.2-Genetic Differences

Figure 1.2 Information Breeding Programs, 08
Figure 1.3 Technological Developments Animal Breeding, 10
Figure 2.1 Decision Areas in Animal Breeding, 14
Figure 2.2 Primary Components Breeding Approach, 17
Figure 2.3 Genetic Differences in a Species, 20
Figure 2.4 Genetic Differences, 22
Figure 2.5 Genetic Differences, 23
Figure 2.6 Environmental Inputs & Genetic Differences, 24
Figure 3.1 Performance Recording in Animal Breeding, 28
Figure 3.2 Performance Recording Interactions, 30
Figure 3.3 Value of Measurement in Animal Breeding, 34
Figure 3.4 Measurement Systems Comparison, 36
Figure 3.5 Information Combination, 38
Figure 3.6 Recordings Required for Effective Monitoring, 43
Figure 4.1 EBV Calculation, 48
Figure 4.2 EBV Estimation, 50
Figure 4.3 Information Recorded & Spread of EBVs, 53
Figure 4.4 Average EBVs of Parents, 54
Figure 5.1 Genetic Pathways Weaning Weight, 67
Figure 5.2 Distributions of Genetic Difffferences, 69
Figure 7.1 McDonald Components of Calving Records, 81
Figure 11.1 Pig Industry Structure, 103
Figure 12.1 Swan Expression of Heterosis, 113
Figure 12.2 Swan Components of Across-Breed Genetic Evaluations, 116
Figure 14.1 Barwick & Fuchs B-OBJECT Features, 154
Figure 15.1 Carrick Selection Index Derivation, 159
Figure 19.1 Kinghorn Relationship Between Offspring & Parent Merit, 194
Figure 19.2 Kinghorn Selection Differentials, 196
Figure 19.3 Kinghorn Selection Intensity, 197
Figure 19.4 Kinghorn Selection Intensity, 198
Figure 19.5 Kinghorn Calculations of Selection Intensity, 199
Figure 19.6 Kinghorn Response to Selection, 200
Figure 19.7 Kinghorn Pattern of Ewe Migration, 202
Figure 19.8 Kinghorn Pattern of Ewe Migration, 203
Figure 20.1 Hammond Operations Contributing Selection Programs, 212
Figure 21.1 Kinghorn Predicted Response in Progeny, 218
Figure 21.2 Kinghorn Artificial Insemination & Selection Intensity, 219
Figure 21.3 Kinghorn MOET & Selection Intensity, 220
Figure 21.4 Kinghorn Merit of Clones & Merit of Progeny, 221
Figure 21.5 Kinghorn Genetic Progress Normal Breeding Program, 223
Figure 21.6 Kinghorn Open Nucleus System with MOET of Females, 224
Figure 24.1 Schneeberger Expected Income Versus Variance Frontier, 245
Fleece Weight Heritability, 49
Flock Structure Form, 163
Form Current Levels of Production, 164
Form Reproduction Rate, 165
Form Sale of Surplus Sheep, 166
Form Sample Breeding Objective Data, 161
Form Target Levels of Production, 168
Form Value of Added Production, 167
Form Value of Production, 166
Form Wool Production, 164
Form the Current Flock Structure, 163
Formal Approach Aims to Maximise Profit Increase, 156
Formulating Breeding Objective, 142
Fuchs W Breeding Objectives for Beef Cattle, 141
Fully Integrated Genetic Evaluation System, 109

GFW Greasy Fleece Weight
GROUP BREEDPLAN EBVs and $INDEX-EBVs, 151
GROUP BREEDPLAN History, 78
Gene Interaction, 19
Generation Turnover Selection Program, 208
Genetic Biases BREEDPLAN, 78
Genetic Change & Breeding Operation, 13
Genetic Change, 07
Genetic Change Direction, 155
Genetic Changes Breeding Program, 18
Genetic Correlation, 157
Genetic Differences, 02, 22, 23
Genetic Differences Extent of, 18
Genetic Differences Manipulating, 193
Genetic Differences in a Species, 20

251

Genetic Effects-Hybrid Vigour

Genetic Effects Direct, 80
Genetic Evaluation, 17
Genetic Evaluation Across-Breed, 111
Genetic Evaluation Commercial Producer, 71
Genetic Evaluation Dairy Industry, 95
Genetic Evaluation Procedures Historical Development, 57
Genetic Evaluation Seedstock Producer, 71
Genetic Evaluation System Fully Integrated, 109
Genetic Evaluation Using BLUP in Sheep, 86
Genetic Evaluation in Meat Sheep, 89
Genetic Evaluation in Wool Sheep, 85
Genetic Evaluation in the Pig Industry, 103
Genetic Evaluation of Beef Cattle, 77
Genetic Evaluation of Calving Ease, 84
Genetic Evaluation of Categorical Traits, 67
Genetic Evaluations Across-Flock, 92
Genetic Evaluations Delivering Through LAMBPLAN, 92
Genetic Evaluations Simple Across-Breed, 114
Genetic Gain with AI & MOET, 217
Genetic Improvement & Product Pricing Systems, 11
Genetic Improvement in Livestock, 01
Genetic Markers, 06
Genetic Parameters Assumed, 147
Genetic Parameters Gestation Length & Weight, 83
Genetic Parameters for Traits in BREEDPLAN, 79, 80
Genetic Prediction Statistical Models, 90
Genetic Progress Normal Breeding Program, 223
Genetic Progress Principles, 193
Genetic Progress in Open Nucleus Schemes, 201
Genetic Resources, 193
Genetic Trends BREEDPLAN, 78
Genetic Trends Comparison, 76
Genetics & Economic Value, 157
Genetics Immunological, 22
Genetics Importance in Animal Breeding, 19
Genetics Industry Use, 171
Genetics Molecular, 22
Genetics Physiological, 22
Genetics Quantitative, 21
Genetics Veterinary, 22
Genotype by Environment Interactions & Analysis, 75
Gestation Length & EBVs, 82
Gianola & Foulley, 68
Goals Defining Pig Industry, 183
Goats Heritability Estimates, 49

Goddard M Breeding Objectives for Dairy Cattle, 177
Goddard M Genetic Evaluation Dairy Industry, 95
Greasy Fleece Weight & Dollar EBVs, 52
Great Britain Breeding Objectives Lambs, 174
Growth Rate Heritability, 104
Growth Traits BREEDPLAN, 79
Guide for Consultants Breeding Objectives for Beef Cattle, 152
Guide for Consultants Breeding Objectives Pigs, 191
Guide for Consultants Crossbreeding Programs, 235
Guide for Consultants Economics and Animal Breeding, 138
Guide for Consultants Genetic Evaluation Pigs, 110
Guide to Consultants Across-Breed Genetic Evaluation, 119

Hammond K Design of Straight-Breeding Programs Common Problems, 205
Hammond K Designing Performance Recording Operations, 27
Hammond K Economic Considerations Animal Breeding, 237
Hammond K Modern Breeding Approach, 13
Hammond K New Era in Genetic Improvement, 01
Hammond K Within Versus Across Herd or Flock Evaluation, 71
Hazel Selection Index, 57
Heifer Selection & EBVs, 99
Henderson, 58
Henderson C BLUP, 06
Henderson's Mixed Model Equations, 64
Herd Life & Milk Production, 99
Herd Size Selection Program, 207
Heritabilities, 121
Heritabilities Carcase Traits, 83
Heritability & EBV, 47
Heritability, 157
Heritability Definition, 195
Heterosis & Breed Effects, 229
Heterosis, 111, 227
Heterosis Expression, 113
Heterotic Effects, 21
History Animal Breeding, 05
History Genetic Evaluation, 77
Horses Heritability Estimates, 49
Hybrid Vigour, 111

Identification Systems, 39
Identification of Sources of Returns &
　Costs in Commercial Herds, 143
Immunological Genetics, 22
Implied Values, 137
Improvement in Livestock Fundamentals, 01
Improvements Due to Cloning, 221
Inbreeding Accumulation, 109
Inbreeding Depression Selection Program, 208
Independent Culling Levels, 122
Index Composition, 150
Index Selection, 122
Index Weightings Derivation, 147
Index Weightings for EBVs, 150
Indices Selection Derivation, 123
Indirect Measurement Traits, 02
Indirect Versus Direct Measures, 35
Industry Consultants & B-OBJECT, 148
Industry Use of Genetics, 171
Information Breeding Programs, 08
Information Combination, 38
Information Flow Breeding Operations, 13
Information Recorded & Spread of EBVs, 53
Information in Animal Breeding, 07
Inheritance Imperfect, 03
Inputs Performance Recording, 29
Integrating Economic Value & Genetics, 157
Interim EBVs or EPDs, 76
Interpretation EBVs, 98

Kinghorn B Design of Crossbreeding
　Programs, 227
Kinghorn B Maximising Improvement
　with AI, MOET & Cloning, 217
Kinghorn B Principles of Estimated
　Breeding Values, 47
Kinghorn B Principles of Genetic
　Progress, 193

LAMBPLAN & Breeding Objectives in
　Meat Sheep, 171
LAMBPLAN, 07, 18
LAMBPLAN Delivering Genetic
　Evaluations, 92
LAMBPLAN Evaluation Structure, 89
LAMBPLAN Predicted Genetic
　Parameters, 172
Labour Requirements & Design of
　Selection Programs, 210
Labour Requirements Selection Programs, 213
Lactation Yield, 95
Likeability & Milk Production, 99

Litter Size Heritability, 49
Litter Weaning Weight Heritability, 49
Livability Heritability, 49
Livestock Genetic Improvement, 01
Loin Eye Area Heritability, 49
Long T Breeding Objectives in the Pig
　Industry, 183
Long T Genetic Evaluation in the Pig
　Industry, 103
Lush, 57

MLC Meat & Livestock Commission
MME Mixed Model Equations
MOET & AI in Open Nucleus Schemes, 222
MOET & Selection Intensity, 220
MOET Increasing Selection Intensity, 217
MOET Maximum Improvement, 217
MSI Multibreed Selection Index
Male Parents Selection Differentials, 196
Management & Crossbreeding, 228
Management Perspectives and Marketing, 135
Management System & Selection
　Program, 209
Management Systems Measurement, 39
Management of Risk, 241
Management the Need for Optimal, 136
Marker Sire Genetic Evaluation, 73
Marker Traits, 35
Markers Molecular, 35
Market Value & Pricing Systems, 11
Marketing Inputs for Terminal Sire Line
　& Maternal Line, 190
Marketing Inputs to $INDEX, 187
Marketing Management and Perspectives, 135
Marketing System Identification, 133, 142
Markets Pigs Breeding Objectives, 183
Markets Winning, 01
Mastitis Susceptibility Heritability, 49
Mate Selection Breeding Program, 18
Maternal Sector Lamb Industry, 173
Mating Decisions, 15
Mating Selection Program, 208
Mature Weight Heritability, 49
Maximising Improvement with MOET
　&Cloning, 217
McDonald A Genetic Evaluation of Beef
　Cattle, 77
Measurement Additional Data, 37
Measurement Direct Versus Indirect, 35
Measurement Identification System, 39
Measurement Labour Requirements, 39
Measurement Locations, 37
Measurement Management Systems, 39

253

Measurement Methodology, 33
Measurement Relevance Performance
 Recording, 32
Measurement Systems Comparison, 36
Measurement Which Animals, 37
Measurement in Animal Breeding, 34
Measurements Performance Recording, 32
Meat & Livestock Commission, 107
Meat Genetic Evaluation, 89
Meat Quality Across-Herd or Flock
 Analyses, 76
Meat Sheep & LAMBPLAN, 171
Meat Sheep Breeding Objectives, 169
Meat Sheep Crossbreeding, 233
Meat Yield Across-Herd or Flock
 Analyses, 76
Mendel G, 04
Mendelian Genetics, 19
Merchandising Results, 16
Merit of Clones & Merit of Progeny, 221
Methods of Evaluation Pigs, 106
Migration of Females in Open Nucleus
 System, 224
Milk Economic Weight Dairy Cows, 179
Milk Production Heritability, 49
Milking Speed & Milk Production, 99
Milking Speed Economic Weight Dairy
 Cows, 179
Milking Speed Heritability, 49
Misunderstandings Animal Breeding, 23
Models Genetic Prediction, 90
Models with Repeated Measures, 62
Modern Animal Breeding Relevant
 Developments, 04
Modern Breeding Approach, 13
Modified Contemporary Comparison, 58
Mohair Production Heritability, 49
Molecular Genetic Markers, 06
Molecular Genetics, 10, 22
Molecular Genetics Misunderstanding, 25
Molecular Markers, 35
Multi-Trait Objective, 125
Multibreed Selection Index, 114
Multiple-Trait Model BLUP, 61

NRM Numerator Relationship Matrix
National Beef Recording Service, 77
New Era Genetic Improvement, 01
New Era in Animal Breeding, 09
New Technology, 09
New Zealand Breeding Objectives Lambs, 174
Non-Genetic Biases BREEDPLAN, 78
Non-linearity of Economic Values, 136

North American Breeding Objectives
 Lambs, 174
Nucleus Breeder Decisions, 183
Nucleus Program Genetic Evaluation, 73
Number Born Heritability, 49
Numerator Relationship Matrix, 62

On-Farm Testing Pigs, 105
On-Farm Tests Advantages &
 Disadvantages, 88
Open Nucleus Schemes AI & MOET, 222
Open Nucleus Schemes Genetic Progress, 201
Optimal Crossbreeding Systems, 231
Overall Type Economic Weight Dairy
 Cows, 179
Overseas Bulls & Breeding Values, 97

PIGBLUP, 17, 18, 108
Pacer Heritability, 49
Pacer Log Earnings Heritability, 49
Pacer Trotter Heritability, 49
Pattern of Ewe Migration, 202
Payoffs Selection Programs, 213
Percent Lean Cuts Heritability, 49
Performance Current & Future, 121
Performance Recording & Artificial
 Breeding, 41
Performance Recording Across-Herd/Flock, 42
Performance Recording Dos & Don'ts, 43
Performance Recording Evolution, 41
Performance Recording Interactions, 30
Performance Recording Operations, 27
Performance Recording Planning, 40
Performance Recording Quality, 41
Performance Recording Systems, 31
Performance Recording Time Allocation, 41
Performance Recording What is It?, 27
Performance Recording Within Herd/Flock, 42
Performance Recording in Animal
 Breeding, 28
Perspectives Management and Marketing, 135
Physiological Genetics, 22
Pig Industry Breeders Competitive
 Position, 184
Pig Industry Breeding Objectives, 183
Pig Industry Defining Goal, 183
Pig Industry Genetic Evaluation, 103
Pig Industry Resources Breeding Program, 184
Pig Industry Structure, 103
Pig Industry Traits of Economic
 Importance, 104
Pigs Across-Breed Evaluation, 118

Pigs-Selection

Pigs Central Testing, 105
Pigs On-Farm Testing, 105
Pigs Per Litter Heritability, 104
Pigs Weaned Per Litter Heritability, 104
Planning Ahead Selection Programs, 207
Planning Performance Recording Program, 40
Polygenetic Traits, 21
Population Genetics, 19
Portfolio of Sire Usage, 242
Post Weaning Gain Heritability, 49
Poultry Heritability Estimates, 49
Practical Application of Crossbreeding
  in Livestock Industries, 232
Predicted Genetic Parameters Using
  LAMBPLAN, 172
Predicted Merit of Progeny, 52
Predicted Response in Progeny, 218
Predictions Genetics Statistical Models, 90
Pricing System Ideal, 11
Pricing Systems & Genetic Improvement, 11
Pricing Systems, 11
Primary Components Breeding Operations, 16
Primary Components Breeding Approach, 17
Principles of Estimated Breeding Values, 47
Principles of Genetic Progress, 193
Problems Design of Straight-Breeding
  Programs, 205
Problems Selection Programs, 209
Processing Cost & Pricing Systems, 11
Product Pricing Systems, 11
Production & Economic Inputs to
  $INDEX, 189
Production Cost & Pricing Systems, 11
Production Current Levels Form, 164
Production Index, 95
Production Inputs to $INDEX, 187
Production System Identification, 133, 142
Production Value Form, 166
Productivity & Pricing Systems, 11
Profit Increase Maximisation, 156
Profitability & Animal Breeding, 11
Profitability Maintaining, 01
Profitability Maximisation, 13
Progeny Selection Differentials, 196
Protein Economic Weight Dairy Cows, 179
Protein Heritability, 49
Pulling Power Heritability, 49

Quantitative Genetics, 21

Raw Product & Pricing Systems, 11
Record Collection & Selection Programs, 211

Record Keeping Selection Programs, 215
Recording Effectiveness, 40
Recording System Selection, 15
Records Selection Program, 208
Regressed Contemporary Comparison, 77
Relative Economic Values, 132
Relativity of Trait Economic Values, 135
Repeatabilities, 121
Replacement Breeding Stock Purchase
  Selection Program, 206
Replacement Stock Breeding Operations, 14
Replacements Risk, 241
Reproduction & LAMBPLAN, 90
Reproduction Rate Form, 165
Reproduction Technologies, 10
Reproduction Traits, 81
Reproductive Rate & Crossbreeding, 228
Restricted Indices, 137
Return on Investment Breeding Programs, 16
Returns & Cost Identification, 133
Returns & Costs Biological Traits
  Influencing, 134
Returns & Costs Biological Traits
  Influencing, 144
Returns & Costs Identification of Sources, 143
Riding Performance Dressage Heritability, 49
Riding Performance Heritability, 49
Riding Performance Jumping Heritability, 49
Risk About Selection of Replacements, 241
Risk Management, 241
Robertson & Rendel, 57
Rotational Crossbreeding, 229
Rotational Crossing, 230

SELIND Software Package Pigs, 185
Sale of Surplus Sheep Form, 166
Schneeberger M Alternative Evaluation
  Procedures, 57
Schneeberger M Management of Risk, 241
Scrotal Circumference Heritability, 49
Scrotal Size & Days of Calving Record, 81
Scrotal Size & Days of Calving, 81
Scrotal Size Heritability, 82
Seedstock Producer, 237
Selecting for More Than One Trait, 137
Selection & Culling, 121
Selection Based on other Sources of
  Information, 200
Selection Criteria, 122
Selection Criteria Dairy Cows Australia, 180
Selection Criteria LAMBPLAN, 90
Selection Decisions, 15
Selection Differentials, 196

255

Selection Index, 122
Selection Index Available Criteria, 146
Selection Index Breeding Program, 18
Selection Index Calculations, 181
Selection Index Derivation, 146, 159
Selection Index Examples, 126
Selection Index Finalisation, 158
Selection Index Pigs, 107
Selection Indices Background & Derivation, 123
Selection Indices LAMBPLAN, 91
Selection Intensity, 197, 198
Selection Intensity Calculations, 199
Selection Intensity Increased, 217
Selection Program & Management System, 209
Selection Program Options, 205
Selection Program Problems, 209
Selection Programs Design, 207
Selection Response, 200
Selection Single Trait Pigs, 106
Selection Tandem, 122
Selection Theory Simple, 193
Selection for Compound Traits, 130
Selection for More than One Trait, 122
Semen Selection & EBVs, 98
Semen Value for Money, 181
Services Per Conception Heritability, 49
Shank Length Heritability, 49
Sheep Across Flock Evaluation, 87
Sheep Across-Breed Evaluation, 118
Sheep Genetic Evaluation Using BLUP, 86
Sheep Heritability Estimates, 49
Sheep Meat Breeding Objectives, 169
Sheep Meat Genetic Evaluation, 89
Sheep Wool Genetic Evaluation, 85
Simple Across-Breed Genetic Evaluations, 114
Simple Genetics, 19
Simple Selection Theory, 193
Single Genes in the Breeding Program, 213
Single Trait Selection Pigs, 106
Single-Trait Model BLUP, 61
Single-Trait Objective, 123
Sire Maternal Grandsire Model BLUP, 60
Sire Model BLUP, 60
Sire Reference Across Flock Evaluation, 87
Sire Reference Evaluation, 87
Sire Referencing, 74
Sire Usage Portfolio, 242
Size Economic Weight Dairy Cows, 179
Solids-Not-Fat Heritability, 49
Sources of Information & Selection, 200
Specific Crosses, 229

Standardisation of Yields, 97
Statistical Analysis Procedures, 10
Statistical Models Genetic Prediction, 90
Steps in Defining Breeding Objective, 133
Stock Identification Selection Programs, 210
Stock Management Selection Programs, 211
Stock Management for Recording, 40
Straight Breeding or Crossbreeding, 15
Straight Versus Crossbreeding Programs, 227
Straight-Breeding Programs Design, 205
Strategic Objectives EBVs, 55
Summary Output Performance Recording, 31
Survival Economic Weight Dairy Cows, 179
Swan A Across-Breed Genetic Evaluation, 111
Swan A Design of Crossbreeding Programs, 227
Swine Heritability Estimates, 49
Synthetic Breeds, 229
Systematic Measuring in Animal Breeding, 27

Table 4.1 Heritability Estimates, 49
Table 4.2 Strategic Objectives EBVs, 55
Table 6.1 Across-Herd or Flock Operations, 73
Table 7.1 McDonald Genetic Parameters for Traits in BREEDPLAN, 79
Table 7.2 McDonald Genetic Parameters for Traits in BREEDPLAN, 80
Table 7.3 McDonald Age Limits BREEDPLAN, 81
Table 7.4 McDonald Genetic Parameters Gestation Length & Weight, 83
Table 7.5 McDonald Heritabilities Carcase Traits, 83
Table 11.1 Long Economically Important Traits & their Heritabilities, 104
Table 12.1 Swan Weightings on Breed & Heterosis Effects, 114
Table 14.1 Barwick & Fuchs First Categorisation of Breeding Situations, 143
Table 14.2 Barwick & Fuchs Economic Value Estimates for Traits, 149
Table 14.3 Barwick & Fuchs Index Weightings for EBVs, 150
Table 14.4 Barwick & Fuchs GROUP BREEDPLAN EBVs and $INDEX-EBVs, 151
Table 14.5 Barwick & Fuchs Comparative Rankings of Young Bulls, 151
Table 16.1 Banks Breed Type Contribution to the Lamb Industry, 170
Table 16.2 Banks Predicted Genetic Parameters Using LAMBPLAN, 172

Table 17.1 Goddard Economic Weights for Breeding Objective, 179
Table 17.2 Goddard Selection Index Calculations, 181
Table 18.1 Long Economic Inputs to $INDEX, 186
Table 18.2 Long Production Inputs to, $INDEX, 187
Table 18.3 Long Marketing Inputs to $INDEX, 187
Table 18.4 Long Economic & Production Inputs to $INDEX, 189
Table 18.5 Long Marketing Inputs for Terminal Sire Line & Maternal Line, 190
Table 18.6 Long Estimated Breeding Values &$EBVs, 190
Table 22.1 Swan & Kinghorn Optimal Crossbreeding Sytems, 231
Table 24.1 Schneeberger Efficient Sets of Portfolios, 244
Tandem Selection, 122
Target Levels of Production Form, 168
Teat Placement Heritability, 49
Technological Developments Animal Breeding, 10
Technology New, 09
Temperament & Milk Production, 99
Temperament Economic Weight Dairy Cows, 179
Test Mating Single Gene Traits, 214
Thoroughbred Racing Heritability, 49
Thoroughbred Racing Time Heritability, 49
Throughbred Racing Log of Earnings Heritability, 49
Time Heritability, 49
Trait Economic Value, 134
Traits Contributing to the Breeding Objective Pigs, 184
Traits Direct Measurement, 02
Traits Economic Value Derivation, 134, 145
Traits Economically Important, 104
Traits Growth BREEDPLAN, 79
Traits Heritability, 104
Traits How Easy to Move, 157
Traits How they Move One Another, 157
Traits Indirect Measurement, 02
Traits Marker, 35
Traits Polygenetic, 21
Traits Reproduction, 81
Traits Room to Move, 158
Traits Selecting for More Than One, 137
Traits Single Gene Breeding Program, 214
Traits of Economic Importance Pig Industry, 104

Traits of the Breeding Objective, 122
Traits other than Production, 99
Transgenic Animals, 07
Trends Breeding Program, 18
Type Score Heritability, 49

Using Economics to Formulate the Breeding Objective, 131

Value for Money in Semen, 181
Value of Added Production Form, 167
Value of Measurement in Animal Breeding, 34
Variability, 158
Veterinary Genetics, 22
Visual Appraisal Pigs, 106

WOOLPLAN, 07
Watson & Crick DNA Structure, 06
Weaning Weight Heritability, 49
Weaning Weight Pigs Heritability, 104
Weight & LAMBPLAN, 90
Weight of Retail Cuts Heritability, 49
Weighting & Standardisation of Yields, 96
Weighting on Breed & Heterosis Effects, 114
Weightings Index Derivation, 147
Western Europe Breeding Objectives Lambs, 174
Withers Height Heritability, 49
Within Flock Evaluation, 85
Within Versus Across Herd or Flock Evaluation, 71
Wool Production, 164
Wool Sheep Breeding Objectives, 155
Wool Sheep Crossbreeding, 234
Wool Sheep Genetic Evaluation, 85
Wool Weight & LAMBPLAN, 90
Woolplan Current, 85
Woolplan Requirements, 85

Yearling Hip Frame Size Heritability, 49
Yearling Weight Heritability, 49
Yield Grade Heritability, 49

B&C
Proudly printed by B&C PRINT & POST
263 Liverpool Street East Sydney N.S.W 2010